Marine Corps Logistics Operations
PSC Box 788400
Twentynine Palms, California 92278-8400

May 6, 2016

FOREWORD

Marine Corps Tactical Publication (MCTP) 3-40B, *Tactical-Level Logistics*, provides the doctrinal basis for the planning and execution of ground and aviation logistic support for Marine air-ground task force (MAGTF) operations at the tactical level of war. It establishes standard terms of reference for tactical logistics and combat service support operations and provides guidance for developing unit standing operating procedures throughout the MAGTF. Marine Corps Tactical Publication 3-40B expands upon MCWP 4-1, *Logistics Operations*, and provides detailed guidance to Marine Corps logisticians for the conduct of tactical-level logistics.

This publication is primarily intended for commanders and their staffs who are responsible for planning and conducting logistic operations support at the tactical level of war. The secondary audience is commanders and staff officers who require logistic support or who will benefit from a greater understanding of this support at the tactical level.

This publication supersedes MCWP 4-11, *Tactical-Level Logistics*, dated 13 June 2000.

Reviewed and approved this date.

M. S. COOK
Colonel, U.S. Marine Corps
Commanding Officer
Marine Corps Logistics Operations Group

Publication Control Number: 147 000011 00

DISTRIBUTION STATEMENT A: Approved for public release; distribution is unlimited.

Tactical-Level Logistics

Table of Contents

Chapter 1   **Fundamentals**

1001	Levels of War	1-1
1002	The Logistic Continuum	1-1
1003	Strategic Logistics	1-2
1004	Operational Logistics	1-2
1005	Tactical Logistics	1-2
1006	Tactical Logistics Versus Combat Service Support	1-3
1007	Marine Corps Organizations	1-3
1008	Logistic Staff Responsibilities	1-4
	a. Joint Task Force	1-4
	b. Marine Forces Component Headquarters	1-7
	c. Marine Air-Ground Task Force	1-7
	d. Combat Service Support Organization	1-8
1009	Functions and Subfunctions of Tactical Logistics	1-8
	a. Supply	1-8
	b. Maintenance	1-8
	c. Transportation	1-8
	d. General Engineering	1-9
	e. Health Services	1-9
	f. Services	1-9
1010	Aviation Logistics Division and Marine Aviation Logistics Squadron	1-13
1011	Tactical Logistic Support External to the Marine Air-Ground Task Force	1-13
	a. Naval Logistics	1-14
	b. Contracting	1-14
1012	Combat Service Support Installations	1-15
	a. Combat Service Support Area	1-15
	b. Beach Support Area	1-15
	c. Landing Zone Support Area	1-15
	d. Repair and Replenishment Point	1-16
	e. Forward Arming and Refueling Point	1-16
	f. Airfields and Air Facilities	1-16
	g. Seaports	1-17

Chapter 2   **Logistic Functional Area Support Operations**

Section I   **Supply**

2101	Logistics Combat Element Supply Support Operations	2-1
	a. Landing Force Supplies	2-1
	b. Sustainment	2-2

	c. Ground Supply Operations During the Amphibious Assault	2-2
	d. Ground Supply Operations During Subsequent Operations Ashore	2-5
2102	Ground Combat Element Supply Support Operations	2-6
	a. Commander's Flexibility	2-6
	b. Supply Trains	2-7
2103	Aviation-Peculiar Supply Support Operations	2-11
	a. Marine Aviation Logistics Squadron	2-11
	b. Replacement Aircraft	2-12
	c. Aircraft Fuel and Ammunition	2-12

Section II   Maintenance

2201	Ground Maintenance Support Operations	2-13
	a. Levels of Maintenance	2-13
	b. Maintenance During the Amphibious Assault	2-14
	c. Maintenance During Transition Periods	2-15
	d. Maintenance During Subsequent Operations	2-15
	e. Organizational Maintenance	2-15
	f. Intermediate Maintenance	2-15
	g. Recovery, Evacuation, and Repair Cycle	2-17
2202	Aviation-Peculiar Maintenance Support Operations	2-18
	a. Marine Aviation Logistics Support Program	2-18
	b. Aviation Logistics Support Ship	2-20
	c. Maritime Prepositioning Ships	2-20

Section III   Transportation

2301	Motor Transport Operations	2-21
	a. Operational Techniques	2-21
	b. Types of Haul	2-21
	c. Hauling Methods	2-22
	d. Cargo Throughput	2-22
	e. Convoy Operations	2-22
	f. Types of Routes	2-23
2302	Port and Terminal Operations	2-23
	a. Ship-to-Shore Movement	2-23
	b. Shore-to-Shore Operation	2-23
	c. Logistics Over-the-Shore	2-23
	d. Joint Logistics Over-the-Shore	2-24
	e. Inland Waterway Operations	2-24
	f. Inland Terminal Operations	2-24
	g. Staging Area Operations	2-24
2303	Aerial Delivery Operations	2-25
2304	Deployment	2-25
	a. External Transportation Agencies	2-25
	b. Modes of Transportation	2-25
2305	Employment	2-26

Section IV	**General Engineering**	
2401	Engineering Tasks	2-28
2402	Marine Air-Ground Task Force Engineering Unit Missions	2-28
	a. Combat Engineer Battalion	2-28
	b. Engineer Support Battalion	2-28
	c. Marine Wing Support Squadron	2-28
2403	Naval Construction Force	2-29
	a. Naval Construction Force	2-29
	b. Naval Construction Regiment	2-30
	c. Naval Mobile Construction Battalion	2-30
	d. Construction Battalion Maintenance Unit	2-30
	e. Amphibious Construction Battalion	2-30
2404	Joint Engineering/ Interagency Engineering	2-30
	a. United States Army Engineering	2-30
	b. United States Air Force Civil Engineering	2-31

Section V	**Health Service Support**	
2501	Role Versus Echelon (Level) of Care	2-31
2502	Marine Air-Ground Task Force Capabilities	2-32
	a. Command Element	2-32
	b. Ground Combat Element	2-32
	c. Aviation Combat Element	2-33
	d. Logistics Combat Element	2-33
	e. Medical Battalion	2-33
	f. Medical Logistics Company	2-33
	g. Dental Battalion	2-34
2503	Capabilities External to the Marine Air-Ground Task Force	2-34
2504	Patient Movement	2-34

Section VI	**Services**	
2601	Combat Service Support Services	2-35
	a. Disbursing	2-35
	b. Postal	2-36
	c. Marine Corps Community Services	2-37
	d. Security Support	2-39
	e. Legal Services	2-39
	f. Civil Affairs Support	2-40
	g. Mortuary Affairs	2-40
	h. Contracting Support Services	2-41
2602	Command Services	2-45
	a. Personnel Administration	2-45
	b. Religious Ministries Support	2-45
	c. Financial Management	2-46
	d. Communications and Information Systems	2-46
	e. Billeting	2-46

		f. Food Service Support	2-46
		g. Band	2-50
		h. Morale, Welfare and Recreation	2-50
Chapter 3	**Command and Control**		
	3001	Establishing Command and Control	3-1
		a. Command and Support Relationships	3-2
		b. Command Relationships	3-4
		c. Support Relationships	3-5
		d. Logistics Combat Element Command Relationships	3-7
		e. Mission Assignments	3-7
	3002	Logistics and Combat Service Support Missions	3-7
		a. Inherent Responsibilities	3-7
		b. Mission Statement Elements	3-9
		c. Standard Missions	3-9
		d. Nonstandard Missions	3-11
	3003	Support Procedures in Tactical Logistics Functional Areas	3-11
		a. Supply	3-12
		b. Maintenance	3-13
		c. Transportation	3-13
		d. General Engineering	3-14
		e. Health Service Support	3-14
		f. Services	3-14
	3004	Command Groups and Control Agencies	3-14
		a. Aviation Ground Support Operations Center	3-14
		b. Ground Combat Element Logistic Operations Center	3-14
		c. Logistics Combat Element Combat Operations Center	3-15
	3005	Movement Control Organizations	3-16
		a. Movement Control Centers	3-17
		b. MAGTF Deployment and Distribution Operation Center	3-17
		c. MAGTF Materiel Distribution Center	3-19
		d. Distribution Liaison Cells	3-19
		e. Terminal Operation Organization	3-19
		f. MAGTF Movement Control Center	3-20
		g. Major Subordinate Command Unit Movement Control Center	3-20
		h. Base Operations Support Group	3-20
		i. Station Operations Support Group	3-20
		j. Flight Ferry Control Center	3-20
	3006	Maritime Prepositioning Force Organizations	3-20
		a. Survey, Liaison, and Reconnaissance Party	3-20
		b. Offload Preparation Party	3-20
		c. Arrival and Assembly Operations Group	3-21

3007	Amphibious Ship-to-Shore Movement Organizations	3-21
	a. Navy Control Organization	3-21
	b. Landing Force Control Organization	3-21
	c. Naval Beach Group	3-23
	d. Other Naval Landing Support Assets	3-24
3008	Communications	3-24
3009	Logistic Information Management	3-24
	a. Organic Capabilities	3-24
	b. Information Systems Functional User Responsibilities	3-25
	c. Information Systems	3-25
3010	Liaison	3-30
	a. Liaison Element	3-30
	b. Liaison Element Selection Considerations	3-30
	c. Exchange of Liaison Element	3-31
	d. Liaison Element Duties and Responsibilities	3-32
	e. Liaison Procedures	3-32

Chapter 4 Planning

4001	Logistic Planning Concepts	4-1
4002	Planning for Expeditionary Operations	4-1
	a. Phases of Action	4-1
	b. Forward-Deployed Logistic Capabilities	4-2
	c. Marine Expeditionary Planning Organization	4-3
4003	Types of Joint Planning	4-4
4004	Marine Corps Planning Process	4-4
	a. Marine Corps Planning Process and the Logistician	4-5
	b. Problem Framing	4-5
	c. Course of Action Development	4-6
	d. Course of Action Wargaming	4-7
	e. Course of Action Comparison and Decision	4-7
	f. Orders Development	4-7
	g. Transition	4-8
4005	Concept of Logistic Support	4-8
4006	Planning Elements	4-8
4007	Planning Techniques	4-9
4008	Deployment Planning Considerations	4-10
4009	Commander's Intent	4-10
4010	Operational Planning Considerations	4-10
	a. Supply	4-10
	b. Maintenance	4-11
	c. Transportation	4-11
	d. External Support	4-11
	e. Forward Support	4-11
	f. Air Support	4-11

	g. Alternate Supply Routes	4-11
	h. Security	4-11
4011	Functional Area Planning Considerations	4-11
	a. Supply	4-11
	b. Maintenance	4-12
	c. Transportation	4-13
	d. General Engineering	4-14
	e. Health Services	4-15
	f. Services	4-16
4012	Coordinating Support	4-16
4013	Intelligence Support	4-17
4014	Host-Nation Support	4-17
4015	Planning Documents	4-18
	a. Logistic/Combat Service Support Estimate	4-18
	b. Annex D to the Marine Air-Ground Task Force Operation Order	4-18
	c. Logistics Combat Element Operation Order	4-19
	d. Standing Operating Procedures	4-19
	e. Other Planning Documents	4-20

Appendices

A	Logistic and Combat Service Support Task-Organization Guide	A-1
B	Sample Format of a Logistic/Combat Service Support Estimate	B-1
C	Sample Format of Annex D (Logistics/Combat Service Support)	C-1
D	Logistic Planning Consideration for MCPP	D-1

Glossary

References and Related Publications

To Our Readers

CHAPTER 1
FUNDAMENTALS

Logistics is defined as "Planning and executing the movement and support of forces." (Joint Publication [JP] 4-0, *Joint Logistics*)

Logistics is a fundamental element of Marine air-ground task force (MAGTF) expeditionary operations. The Marine Corps provides self-contained and self-sustained expeditionary forces designed to independently accomplish missions. These forces are task-organized to meet a wide range of missions and have the logistic capabilities to initiate an operation, to sustain and to reconstitute the forces for follow-on missions.

This publication considers logistic support from the perspectives of supported (e.g., ground combat element [GCE]) and supporting (e.g., logistics combat element [LCE]) organizations. Effective logistics emphasizes detailed planning and close integration of the logistic capabilities and capacities of both the supported units and supporting units.

Combat service support (CSS) is the essential capabilities, functions, activities, and tasks necessary to sustain all elements of operating forces in theater at all levels of war. Combat service support includes, but is not limited, to supply, maintenance, transportation, general engineering, health services, and other services required by aviation and ground combat forces to permit those units to accomplish their missions.

1001. LEVELS OF WAR

Military operations require specific logistic support that is based on the strategic, operational, or tactical levels of war. The strategic level of war determines national or multinational (alliance or coalition) strategic security objectives and guidance, then develops and uses national resources to achieve those objectives. Campaigns and major operations are planned, conducted, and sustained to achieve strategic objectives within theaters or other operational areas at the operational level of war. At the tactical level of war, battles and engagements are planned and executed to achieve military objectives assigned to tactical units or task forces.

1002. THE LOGISTIC CONTINUUM

Strategic, operational, and tactical logistics parallel and complement the levels of war. Strategic logistics supports the organizing, training, and equipping of forces needed to further the national interest. Operational logistics links tactical requirements and strategic capabilities to accomplish operational goals and objectives. Tactical logistics includes organic unit capabilities and combat service support activities required to support military operations.

Effective tactical logistics support results from the proper employment of logistic capabilities within the MAGTF concept of operations and scheme of maneuver. Commanders and logisticians should carefully integrate logistic considerations into operations planning and execution. Tactical-level logistic capabilities are a primary element of a self-sufficient MAGTF, which is supported externally through logistic activity at the strategic and operational levels. Figure 1-1 depicts the continuum of logistic support through the levels of war.

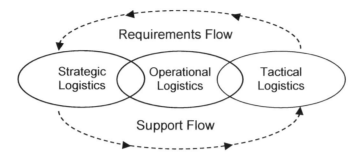

Figure 1-1. The Logistic Continuum.

1003. STRATEGIC LOGISTICS

Strategic logistic capabilities are generated based on guidance from either the President or the Secretary of Defense and logistic requirements identified by the operating forces. The combatant command plans and oversees logistics from a theater-strategic perspective. The joint staff and combatant commanders generate and move forces and materiel into theater and areas of operations where operational logistic concepts are employed.

1004. OPERATIONAL LOGISTICS

Operational logistics connects the logistical activities of the strategic level with those of the tactical level. The Marine Corps is responsible for operational logistic support systems and platforms, and their execution, including life-cycle readiness. Within a theater of operations, the Marine component commander is responsible for conducting operational logistics and coordinating operational logistic support with tactical logistic operations. While the Marine Corps is not normally tasked to execute operational-level logistics, the Marine Corps component commander may be augmented and/or may task elements of a logistics combat element (LCE) to perform operational-level functions. Integration with strategic-level support is coordinated through the Marine Corps component commander.

1005. TACTICAL LOGISTICS

The MAGTF commander plans and executes tactical logistics while coordinating with higher headquarters (HHQ) for operational-level logistic support to sustain operations. Subordinate element commanders within the MAGTF are responsible for the efficient employment of organic logistic capabilities, while the LCE commander is also responsible for executing CSS operations in support of the entire MAGTF.

All elements of the MAGTF execute tactical logistics to some degree by employing organic capabilities. The initial source of logistical support available to any unit is its own organic capabilities further defined in unit tables of organization (TO) and tables of equipment (TE). The LCE possesses capabilities beyond those found in the other MAGTF elements, and provides additional logistical support to the other elements require.

1006. TACTICAL LOGISTICS VERSUS COMBAT SERVICE SUPPORT

As explained in chapter 1 of Marine Corps Doctrinal Publication (MCDP) 4, *Logistics*, the terms "logistics" and "combat service support" are often used interchangeably. However, logistics "…encompasses all actions required to move and maintain forces. This includes the acquisition and positioning of resources as well as the delivery of those resources to the forces." Combat service support, on the other hand, "…is the *activity* which actually provides services and supplies to the combat forces." As the majority of the Marine Corps' focus is at the tactical level of war, "combat service support has been considered to be essentially the same as tactical logistics. Indeed, Marine tactical units have logistic officers and logistic sections, but the units that perform logistic functions for these units are referred to as combat service support elements." Further confusing the issue, some combat service support occurs at the operational and perhaps even strategic levels. For example, health services are provided at the operational levels through the use of hospital ships, fleet hospitals, and permanent military medical facilities.

1007. MARINE CORPS ORGANIZATIONS

Commanders, staff officers, and logisticians at all levels should understand the logistics and CSS capabilities of the MAGTF in order to plan effectively for the tactical phases of expeditionary operations. See figure 1-2. In addition, Marine Corps Reference Publication (MCRP) 5-12D, *Organization of the United States Marine Corps,* provides an in depth view and guide to defining structure, command relationships, and mission of Marine Corps units.

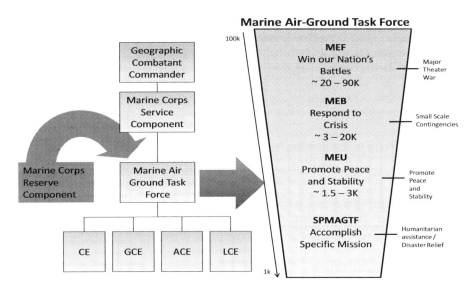

Figure 1-2. Marine Air-Ground Task Force.

1008. Logistic Staff Responsibilities

The logistic staff officer is the commander's principal assistant for logistics and the focal point for policy formation and overall logistical coordination within the organization as well as supported and/or supporting commands. These officers also initiate and maintain continuous liaison with other organizational elements, HHQ, other Services, and allied forces throughout the planning and execution of military operations. See table 1-1 on page 1-5 and table 1-2 on page 1-6.

a. Joint Task Force

Normally, MAGTFs operate as part of a joint or combined task force. A Marine expeditionary force (MEF) or Marine expeditionary brigade (MEB) may serve as the nucleus for such a task force, especially when a Marine expeditionary unit (MEU) is already in theater as the result of forward deployment. In such cases, the Marine Corps component commander may be tasked to provide the joint task force (JTF) headquarters (HQ) nucleus; the MEU may become the initial logistic capability on site. The JTF commander requires direct connectivity with the combatant commander (CCDR) and with the entire JTF. Work with non-DOD, international and local agencies, as well as all components of the JTF requires enhanced command and control (C2), liaison, and support for logistics. The MAGTF G-4/S-4 may become the J-4 for the JTF and perform the following functions:

- Formulate logistic plans.
- Coordinate and supervise:
    - Supply.
    - Maintenance.
    - Repair.
    - Evacuation.
    - Transportation.
    - Engineering.
    - Salvage.
    - Procurement.
    - Operational contract support (OCS).
    - Health services.
    - Mortuary affairs.
    - Communications systems.
    - Host-nation support (HNS).
    - Other related logistic activities.
- Understand the established policies of the other military Services operating as part of the JTF.
- Advise the commander of the logistical support that can be provided for proposed courses of action (COAs).
- Formulate policies to ensure effective logistic support for all forces in the command.
- Coordinate the execution of the commander's policies and guidance.
- Establish a multinational logistic support element to coordinate multinational logistical operations.

Table 1-1. Officers Responsible for CE, ACE, GCE, and LCE Logistics.

General Staff	Chief of Staff	Manpower or Personnel Staff Officer	Operations Staff Officer	Logistic Staff Officer	Aviation Logistic Officer	Comptroller
Special staff officer (logistics)	Provost marshal Staff judge advocate	Adjutant personnel officer Morale, welfare, and recreation officer Postal officer Disbursing officer Llegal officer	Civil affairs officer LCE - Ground supply support coordinator - Ground maintenance support coordinator - Transportation support coordinator - Engineer support coordinator - Medical support officer - Dental support officer - Support officers for services functions	Gground supply officer Aviation supply officer Fiscal officer Maintenance man-agreement (ground equipment) officer Ordnance officer Aviation ordnance officer Engineer Airfield services officer Motor transport officer Strategic mobility officer Embarkation officer Surgeon (medical) Dental officer Food services officer Contracting officer	Aviation supply officer Aviation maintenance officer Aviation ordnance officer avionics officer	Disbursing officer Fiscal officer USN budget and accounting officer USMC budget and accounting officer

1. Individual commands may vary based on the commander's preference and/or availability of personnel.
2. Normally, staff structure at lower levels parallels staff structure at the element level.
3. Aviation logistic, supply, maintenance, ordnance, and avionics officers are unique to ACE headquarters. In ACEs based on a single aircraft group or composite squadron, these posts are normally assumed as additional duties by the commanding officer of the assigned host MALS and the squadron or detachment staff.
4. The staff judge advocate and the legal officer coordinate legal functions within the command and between the command and the LCE legal services support section.
5. If the command does not have a comptroller the disbursing officer or fiscal officer assumes the comptroller's duties.
6. In the LCE, the G-3/S-3, through functional-area support officers, is responsible for ground CSS operations in support of the MAGTF. The LCE G-3/S-3 normally does not supersede the cognizant staff officers (e.g., G-1/S-1, G-4/S-4, etc.) for internal support of the LCE.
7. The aviation ordnance officer and strategic mobility officer are assigned to MEF common equipment facilities.
8. The supply officer, under the cognizance of the G-4/S-4, may also be designated as the fiscal officer.
9. The USMC and USN budget and accounting officers are unique to the ACE.
10. Contracting officers are only available at the CE and LCE.

Table 1-2. CE, ACE, GCE, and LCE Tactical-Level Logistic Responsibilities.

General Staff	Chief of Staff	Manpower or Personnel Staff Officer	Operations Staff Officer	Logistic Staff Officer	Aviation Logistic Officer	Command, Control, Communications, and Computers Systems Officer	Comptroller
Supply				Ground supply (aviation supply)	Aviation supply Aviation ordnance		
Maintenance				Ground maintenance	Aviation maintenance Avionics Aviation ordnance		
Transportation				Transportation			
General Engineering				General engineering			
Health Services				Force health protection First responder capabilities Forward resuscitative care Dental services			
Services	**CSS Services**						
	Security Legal	Disbursing Postal Exchange Legal	Civil affairs LCE - Disbursing - Postal - Exchange - Security - Legal services - Mortuary affairs				
	Command Services						
	Religious Ministries	Band Personnel administration Morale, welfare, and recreation		Financial management Billeting Messing		Communications and information services	Financial management

1. Individual commands may vary based on the commander's preference and/or availability of personnel.
2. Normally, staff structure at lower levels parallels staff structure at the element level. However, at lower levels special staff responsibilities may be assigned as additional duties rather than as primary duties.
3. The aviation logistic officer is unique to ACE and MAW headquarters. In ACEs based on a single aircraft group or composite squadron, this posts is normally assumed as additional duties by the commanding officer of the assigned host MALS and the squadron or detachment staff.
4. The staff judge advocate and the legal officer coordinate legal functions within the command and between the command and the LCE legal services support section.
5. In the LCE, the G-3/S-3, through functional-area support officers, is responsible for ground CSS operations in support of the MAGTF. The LCE G-3/S-3 normally does not supersede the cognizant staff officers (e.g., G-1/S-1, G-4/S-4, etc.) for internal support of the LCE.
6. At a MEF common equipment facility, the logistic officer is responsible for aviation supply.
7. The logistic officer is responsible for financial management if the command does not have a comptroller.
8. The supply officer, under the cognizance of the G-4/S-4, may also be designated the fiscal officer.

b. Marine Forces Component Headquarters

When operating ashore, Marine forces are usually part of a joint or combined force, and the commander, Marine Corps forces (COMMARFOR) is the component command under the geographic combatant commander. The MAGTF commander may serve as COMMARFOR and must comply with operational direction from the joint force commander (JFC). The COMMARFOR must be capable of coordinating combat, combat support, and CSS activity with adjacent units from other Services and allied nations as well as exercising operational control over assigned forces. Consequently, the MAGTF G-4/S-4 must be able to execute operational logistical functions. The COMMARFOR G-4 is responsible for the following functions:

- Advising the commander and operations staff officer (G-3) on the support required to sustain campaigns and major operations.
- Identifying requirements and coordinating the distribution of resources within the strategic infrastructure.
- Anticipating tactical logistic requirements.
- Maximizing the overall effect of support so that the deployment and employment of the force are balanced.
- Planning and supervising the establishment and operation of intermediate and forward support bases.
- Supervising the reception, staging, on-ward movement, and integration of Marines reaching the theater.
- Coordinating with joint, other Service, and host nation agencies for logistic support.
- Planning and supervising the reconstitution and redeployment of the MAGTF for follow-on missions.
- Contract planning, integration, and synchronization with all OCS matters.

c. Marine Air-Ground Task Force

The MAGTF G-4/S-4 is responsible for the following functions:

- Advising the commander and the G-3/S-3 on the readiness status of major equipment and weapons systems.
- Developing policies and identifying requirements, priorities, and allocations for logistical support.
- Integrating organic logistic operations with logistical support from external commands.
- Coordinating and preparing the logistic and CSS portions of plans and orders.
- Supervising the execution of the commander's orders regarding logistics and CSS.
- Ensuring that the logistic support concept supports the overall concept of operations and the scheme of maneuver by identifying and resolving support deficiencies.
- Collating the support requirements of subordinate organizations by identifying the support requirements that can be satisfied with organic resources and passing unsatisfied requirements to the appropriate higher and/or external command.
- Supervising some command services, such as messing and, as directed, billeting and financial management functions.

- Coordinating with the amphibious task force N-4 (if applicable) and the ACE G-4/S-4 for aviation specific requirements.
- Contract planning, integration and synchronization with all OCS matters.

d. Combat Service Support Organization

The ground-common or aviation-peculiar, logistic support CSS organization G-3/S-3 coordinates with supported organizations for their support requirements. The G-3/S-3 is responsible for—

- Coordinating with both the G-3/S-3 and G-4/S-4 of the supported organizations to identify support requirements and to develop estimates of supportability for their concepts of operations.
- Recommending the task organization of supporting LCEs based on guidance from HHQ, the concept of operation, and schemes of maneuver of the supported organizations.
- Coordinating and supervising execution of the command's logistical support operations and providing liaison elements to the supported commands. (The LCE is the primary agency for nonaviation peculiar logistical support operations in the MAGTF and the ACE is responsible for aviation peculiar support.)
- Coordinating with the G-3/S-3 of the supported organizations during the development of their concepts of operations and schemes of maneuver to ensure that they are supportable.

1009. FUNCTIONS AND SUBFUNCTIONS OF TACTICAL LOGISTICS

Marine Corps tactical logistics is categorized in six functional areas: supply, maintenance, transportation, general engineering, health services, and services. See table 1-3 on page 1-10.

a. Supply

Supply involves the requisition authority, distribution, care of supplies while in storage, and salvage of supplies, including the determination of kind and quantity of supplies. See table 1-4 on page 1-11. Logisticians normally calculate requirements for each class and subclass of supply. For additional guidance governing the principles of concepts of supply, as well as the organization, integration, planning, and execution of MAGTF expeditionary supply support, see Marine Corps Warfighting Publication (MCWP) 4-11.7, *MAGTF Supply Operations*.

b. Maintenance

Maintenance involves those actions taken to keep materiel in serviceable condition (preventive maintenance) and actions required to return materiel to serviceable condition (corrective maintenance). Maintenance tasks are grouped by levels of support that determine assignment of maintenance responsibilities. Tables 1-5 and 1-6, on page 1-12, depict the levels of support as they are defined for ground equipment and aviation equipment, respectively; tactical logistic maintenance levels are highlighted. See MCWP 4-11.4, *Maintenance Operations*, for additional information.

c. Transportation

Transportation is moving from one location to another using railways, highways, waterways, pipelines, oceans, and airways. Throughput is the amount of cargo and personnel passed through the transportation systems. The transportation system includes the means and the controls for managing the transportation methods. The transportation subfunctions are generally applicable to all levels of support, although the means, methods, control, and management procedures employed at each level will vary. For more information on transportation operations, see MCWP 4-11.3, *Transportation Operations.*

d. General Engineering

General engineering is considered a tactical logistic function, while combat engineering is considered a combat support function. The MAGTF may receive additional augmentation from the naval construction force (NCF) based on mission, enemy, terrain and weather, troops and support available–time available (METT-T), and space, and logistics; and the size/scope of general engineering effort required. The NCF can range in size and capability from a naval construction regiment (NCR) to a detachment from a naval mobile construction battalion (NMCB) or construction battalion maintenance unit (CBMU). Engineer support battalion (ESB) assets at the tactical level may be used to reinforce or augment the combat engineer battalion (CEB) or Marine wing support squadron (MWSS) with engineering specific capabilities for mobility, countermobility, or survivability tasks. These assets are normally in general support of the MAGTF for a wide range of tasks. These tasks often involve more detailed planning and preparation and higher standards of design and construction than typical combat engineer tasks. For more information on the engineering organizations, functions, and capabilities, see MCWP 3-17.7, *General Engineering,* and MCWP 4-11.5, *Seabee Operations in the MAGTF.*

e. Health Services

Health service support (HSS) refers to those activities and organizations that minimize the effects that wounds, injuries, and disease have on unit effectiveness, readiness and morale. The mission is accomplished by an aggressive and proactive preventative medicine program that safeguards personnel against potential health risks and by establishing an HSS system that provides appropriate care from the point of injury/illness to the appropriate taxonomy of care. For more information on HSS organizations, functions, and capabilities, see MCWP 4-11.1, *Health Service Support Operations.*

f. Services

Within the Marine Corps, services are considered to be one of the six functional areas of tactical logistics. The Marine Corps subdivides the services functional area into two components: command support and combat service support services.

Command support is the responsibility of the commander. Just as the commander is legally liable for the public funds expended by his organization, the commander is equally responsible for the health, welfare, and morale of the personnel in his command. Services that are command support are inherent in any organization that resides within the headquarters element of each unit of the MAGTF. These services include the following: personnel administration; religious ministry; billeting; financial management; food service support; and morale, welfare, and recreation (MWR). Combat service support services are primarily provided by two organizations of the MAGTF: the aviation combat element (ACE) and the LCE. Marine air-ground task force operations require the LCE to be able to provide logistical support services to both mobile and stationary organizations (such as logistic support bases/base camps). In the ACE, the MWSS provides limited services to address ACE requirements.

Other support services may be provided by other organizations such as law enforcement battalion (e.g., security support services). The LCE provides services including OCS, postal, legal, mortuary affairs, security support, civil-military operations, and hygiene/waste management. For more information, see MCWP 4-11.8, *Services in an Expeditionary Environment.*

Table 1-3. Functions and Subfunctions of Tactical Logistics.

Supply	Maintenance	Transportation
<ul><li>Determination of requirements</li><li>Requisition authority</li><li>Storage</li><li>Procurement</li><li>Distribution</li><li>Salvage</li><li>Disposal</li></ul>	<ul><li>Inspection and classification</li><li>Servicing and repair</li><li>Modification</li><li>Rebuilding and overhaul</li><li>Reclamation</li><li>Recovery and evacuation</li></ul>	<ul><li>Embarkation and landing support</li><li>Port and terminal operations</li><li>Motor transport</li><li>Aerial delivery</li><li>Freight and passenger transportation</li><li>Material handling equipment</li></ul>
General Engineering	**Health Services Support**	**Services**
<ul><li>Engineer reconnaissance</li><li>Bridging</li><li>Horizontal/vertical construction</li><li>Facilities maintenance</li><li>Demolition/obstacle removal</li><li>Explosive ordnance disposal</li><li>Receive, store, and distribute bulk fuel</li><li>Water production and storage</li><li>Power generation and distribution</li></ul>	<ul><li>Casualty management</li><li>Force health protection and prevention</li><li>Medical logistics</li><li>Medical command and control</li><li>Medical stability operations</li></ul>	**Command:**<ul><li>Personnel administration</li><li>Religious ministries support</li><li>Financial management</li><li>Communications</li><li>Billeting</li><li>Food service and subsistence support</li><li>band</li><li>Morale, eelfare and recreation</li></ul>**CSS:**<ul><li>Disbursing</li><li>Postal</li><li>MCCS exchange services</li><li>Security</li><li>Legal services</li><li>Civil affairs</li><li>Mortuary affairs</li><li>Contracting</li></ul>

Table 1-4. Classes of Supply.

Class	Description	Subclass
I	Subsistence, which includes rations and gratuitous health and welfare items	A–air (in-flight rations), C–operational rations, R–refrigerated subsistence, S–nonrefrigerated, W–ground
II	Minor end items, which include: clothing, individual equipment, tentage, organizational tool sets and tool kits, hand tools, maps, administrative and housekeeping supplies and equipment	A–air, B–troop support materiel, E–general supplies, F–clothing and textiles, H–test, measurement and diagnostic equipment, W–ground, Z–chemical
III	Petroleum, oils, and lubricants, which include petroleum fuels, lubricants, hydraulic and insulating oils, preservatives, liquid and compressed gases, bulk chemical products, coolants, de-icing and antifreeze compounds and the components and additives of such products, and coal	A–air, W–ground (surface), 1–air bulk fuels, 2–air packaged bulk fuels, 3–air packaged petroleum products, 4–ground bulk fuels, 5–ground packaged bulk fuels, 6–ground packaged petroleum, and 7–ground solid fuels
IV	Construction, which includes construction materiel, installed equipment, and all fortification or barrier materiel	T–industrial supplies
V	Ammunition of all types, which includes chemical, biological, radiological, and special weapons, bombs, explosives, mines, fuzes, detonators, pyrotechnics, missiles, rockets, propellants, and other associated items	A–air, L–missiles, N–special weapons, and W–ground
VI	Personal demand items and nonmilitary sales items	
VII	Major end items, which are the final combination of end products assembled and configured in their intended form and ready for use (e.g., launchers, tanks, mobile machine shops, vehicles)	A–air, B–troop support materiel (includes power generators and construction, barrier, bridging, fire-fighting, petroleum, and mapping equipment), D–commercial vehicles, G–electronics, K–tactical vehicles, L–missiles, M–weapons, N–special weapons, O–combat vehicles, Q–marine equipment, U–communication security materiel, W–ground, Y–railway equipment, and Z–chemical
VIII	Medical materiel, which includes supplies to support health service support, force health protection and medical-unique repair parts	A–medical and/or dental materiel and B–blood and blood products
IX	Repair parts, which include components and kits, assemblies, and subassemblies (reparable and nonreparable) required for maintenance support of all equipment	A–air, L–missiles, N–special weapons, T–industrial supplies, W–ground, and Z–chemical
X	Nonmilitary materiel, which includes materiel to support nonmilitary programs (e.g., agriculture, economic development), that is not included in classes I-IX	

Table 1-5. Levels of Maintenance.

Levels of Maintenance	Categories of Maintenance
Field. Field maintenance is performed by crew/operators and maintainers within Marine Corps organizations and activities, and/or by approved commercial/contract sources. Maintenance tasks performed within the field levels of maintenance are categorized as organizational and intermediate. A unit may perform any field maintenance tasks for which it is manned, trained, and equipped.	Organizational—Authorized at, performed by, and the responsibility of the using unit. Consists of cleaning, servicing, inspecting, lubricating, adjusting, and minor repair. Intermediate—Performed by designated agencies in support of the using unit or, for certain items of equipment, by specially authorized using units. Includes repair of secondary reparables, subassemblies, assemblies, and major end items for return to lower levels or to supply channels.
Depot. Maintenance actions taken on materiel or software involving the inspection, repair, overhaul, or the modification or reclamation (as necessary) of weapons systems, equipment end items, parts, components, assemblies, and subassemblies that are beyond field maintenance capabilities.	Depot maintenance is an essential part of supporting and/or extending equipment lifecycle in total lifecycle management, but may also be leveraged to contribute to field maintenance efforts by providing overflow, on-site maintenance services, and technical assistance as appropriate to maintain enterprise materiel availability.

Table 1-6. Levels of Aviation Equipment Maintenance Activities.

Levels of Maintenance	Maintenance Activities
Organizational	Tactical and training squadrons and Marine Corps air stations with aircraft assigned
Intermediate	Marine aviation logistics squadrons (MALS)
Depot	Naval aviation depots and contract maintenance depot activities Each MALS has limited depot-level capability

1010. AVIATION LOGISTICS DIVISION AND MARINE AVIATION LOGISTICS SQUADRON

There are two combatant command-level Marine Corps component commands with standing aviation logistics divisions (ALDs): MARFORCOM and MARFORPAC. All other combatant command-level Marine Corps component commands can be augmented from MARFORCOM, MARFORPAC, and MAW ALD staffs to provide ALD staff functionality as required. The assistant chief of staff ALD as a primary staff branch of a Marine Corps component command headquarters is responsible for strategic and operational aspects of aviation logistics (AVLOG) for forces assigned under their cognizance. The assistant chief of staff ALD is responsible for—

- Advising Marine component commanders on readiness, policies, deliberate planning, organization, functions, and operations.
- Reviewing and assisting in preparation or revision of weapon systems planning documents, program planning documents, and other long-range AVLOG planning tools.
- Managing, distributing, and maintaining accountability of mobile facilities and ancillary equipment.
- Providing AVLOG assistance as required to bases and stations.
- Managing the special functional areas of the ALD: aircraft maintenance, aviation supply, avionics, ordnance, aviation logistic information management and support, and future operations.

The Marine aircraft wing ALD assists subordinate Marine aircraft groups (MAGs) in matters related to aviation material readiness and internal material management of weapon systems and advises the commander on all AVLOG matters. The ALD's goals are to maintain high aircraft and system readiness, minimize costs associated with maintaining aircraft, and improve AVLOG efficiency. This is accomplished through close coordination with HHQ, supporting naval and commercial organizations, and subordinate commands. The ALD organization consists of six core functional branches: aircraft maintenance, aviation supply, avionics, aviation ordnance, AVLOG plans, and aviation logistic information management and support. At the MEF level, the MAGTF will be staffed with an aviation ordnance capability only and all other functions of AVLOG will be managed by the ACE. However, the AVLOG function on a MEB will be coordinated by the subordinate MALS staff. The AVLOG function on a MEU is absorbed between the ACE commander and subordinate staff, and the supporting intermediate-level support facility.

1011. TACTICAL LOGISTIC SUPPORT EXTERNAL TO THE MARINE AIR-GROUND TASK FORCE

Cross-Service support is appropriate when there are standing Department of Defense (DOD) procedures for common-item support (e.g., for materiel managed by the Defense Logistics Agency) or there are existing inter-Service support agreements (e.g., for the US Army to provide line-haul transportation to Marine forces [MARFORs] in certain theaters). Commanders of unified commands have directive authority for logistics by which they may authorize cross-Service support within their theater. Coalition, bilateral, and/or HNSt agreements authorize specified support across national lines. Requests for cross-Service or cross-national logistic support are coordinated by the Marine component commander.

a. Naval Logistics

The implementation/utilization of Naval Logistics Integration (NLI) provides a naval logistic capability that operates afloat or ashore and with a similar footprint and improved responsiveness than separate logistic chains maintained by each Service. The Secretary of the Navy Instruction (SECNAVINST) 4000.37A, *Naval Logistics Integration*, directs that all levels of command pursue integrating Service logistical capabilities across the Navy and Marine Corps. See MCWP 4-2, *Naval Logistics,* and the NLI Playbook for additional information about operational and tactical-level mission areas, enabling functions, and the organization and support for the conduct of logistic operations for naval Services.

The multi-tiered NLI organization encompasses Navy and Marine Corps organizations across the full spectrum of logistic planning and execution. The NLI strategic plan provides the NLI roadmap (goals and objectives), while SECNAVINST 4000.37A codifies the strategic plan and assigns Service responsibilities for implementing NLI. Subsequently published, the NLI Playbook establishes a common set of tactics, techniques, and procedures for leveraging NLI sanctioned initiatives in support of naval expeditionary forces.

b. Contracting

Contracting personnel at the MARFOR and MEF levels will provide overarching guidance, through OCS, on how to best use, leverage, integrate, synchronize, and manage contracting support in the operational environment. For more information on external contracting resources, see Department of Defense Directive (DODD) 2010.9. In addition to organic contracting capabilities, the MAGTF commander has numerous other support resources available to assist in providing logistic support as described below:

- Acquisition and cross-servicing agreements (ACSAs) are agreements that are allowed under DODD 2010.9 and give MAGTF commanders the authority to acquire or provide logistical support, supplies, and services directly from/to eligible countries and international organizations. The supplies/services available depend on the specific agreement between the United States and the partnering nation. Listings of ACSAs and the supplies and services that are specifically allowed and how they are to be accounted for are available via the ACSA Global Automated Tracking and Reporting System. Each MARFOR has specific policies and procedures regarding ACSA.
- A status of forces agreement (SOFA) is an agreement between a host country and a foreign nation stationing forces in that country. The SOFA is intended to clarify the terms under which the foreign military is allowed to operate. Items addressed by SOFAs may include issues like entry and exit into the country, tax liabilities, postal services, and employment terms for host-country nationals.
- HNS agreements are typically used in nations in which a SOFA does not exist due to the fact that there are no foreign forces stationed in the host country, but in which there is sufficient presence to warrant a basic agreement delineating the rights/responsibilities of the host nation (HN) and foreign military that is visiting.

- Logistic civil augmentation program/Air Force contract augmentation program may be authorized by the lead agency during joint operations. The objective of these programs is to preplan for the use of civilian contractors to perform selected services in wartime to augment Army forces (or all forces during joint operations). Nongovernmental organizations are private, self-governing, not-for-profit organizations dedicated to alleviating human suffering and/or promoting education, healthcare, economic development, environmental protection, and human rights and/or encouraging the establishment of democratic institutions and civil society. Marine air-ground task force commanders can work in coordination with a nongovernment organization and may be able to leverage direct support or supplies from it.

1012. COMBAT SERVICE SUPPORT INSTALLATIONS

The LCE establishes fixed installations to build up logistic capabilities. Installations are physical locations, either aboard ship or ashore, in support of the MAGTF. Their number, location, and specific capabilities are dictated by the concept of CSS, which is based on the MAGTF mission and concept of operation. The MAGTF concept of operation should address the requirement to defend and protect installations and facilities, as required.

a. Combat Service Support Area

The LCEs operate combat service support areas (CSSAs) in accordance with the MAGTF Concept of Logistics Support and the LCE operation order (OPORD). The CSSAs are often designated as primary targets by enemy forces, which directly affects the MAGTF's ability to sustain operational tempo. The landing force (LF) must ensure their security by integrating CSSAs into ground defense and fire support plans and by employing dispersion. The beach support area (BSA) or landing zone support areas are often developed into CSSAs when the LCE establishes the necessary CSS capabilities in the installation to support sustained operations.

b. Beach Support Area

In amphibious operations, the BSA is the area to the rear of a landing force, that contains the facilities for the unloading of troops and materiel for the support of the forces ashore.

The BSA is one of the first CSS installations established ashore during an amphibious operation and maritime prepositioning force (MPF) operations involving in-stream offload. It is established by the shore party group or team, but the LCE commander may eventually disestablish it or consolidate it as part of the CSSA. In some situations, the BSA may be the only CSS installation ashore; in other situations, it may be one of several CSS installations.

c. Landing Zone Support Area

The landing zone support area is a forward support installation that provides minimum essential support to the air assault forces of the MAGTF. It can expand into a combat service support area but it is most often a short term installation with limited capabilities, normally containing dumps for rations, fuel, ammunition, and water only; maintenance is generally limited to contact teams and/or support teams.

This CSS installation is established to support assault support elements. It is established by the LCE when a buildup of supplies or other CSS capabilities is anticipated. When a logistic buildup is not planned, the supported unit is responsible for helicopter support team (HST) operations associated with support of the air assault force.

d. Repair and Replenishment Point

A repair and replenishment point is a combat service support installation, normally in forward areas near the supported unit, established to support a mechanized or other rapidly moving force. It may be either a prearranged point or a hastily selected point to rearm, refuel, or provide repair services to the supported force. Depending on the size of the supported force, the LCE may establish multiple points.

Although the main body of the LCE normally follows in trace of the advancing mechanized force, repair and replenishment points are normally in forward areas near the supported unit. This presents some unique C2 problems because CSS assets can become scattered over a wide area. The LCE can also select repair and replenishment points farther to the rear of the mechanized force. Optimally, however, the CSS unit minimizes handling of supplies by having vehicles from the rear make deliveries directly to the users at repair and replenishment points.

e. Forward Arming and Refueling Point

A forward arming and refueling point (FARP) is a temporary facility organized, equipped, and deployed to provide fuel and ammunition necessary for the employment of aviation maneuver units in combat. "The objective at the FARP is to minimize response time and decrease turnaround time in support of combat operations." (MCWP 3-21.1, *Aviation Ground Support*) The ACE commander may establish a FARP to support the operating force scheme of maneuver.

The FARP locations are selected where natural camouflage and terrain features can hide equipment and aircraft. Good drainage and room for tactical dispersion (helicopter servicing, fueling, arming) are of primary importance. Towns and villages are usually ideal locations because they provide hard land for easy movement of aircraft and wheeled vehicles, intersecting road networks, and excellent night operation capabilities.

After selection of the site, preloaded supplies (e.g., refueling equipment, bladders, ammunition) can be transported to the site by truck along with material handling equipment and personnel. Assault support may be used for rapid, initial emplacement of the FARP. Resupply may be accomplished by air or surface transportation. Under certain situations, a combination of aerial and ground-established FARPs may be operationally desirable. The FARPS are usually established in or near the forward assembly areas. Locations and routes to and from FARPs should be masked from radar detection. Because of the volume of air traffic and its importance to assault support operations, FARPs should be kept beyond medium artillery range. To minimize this threat, FARPs must be displaced often when they are located farther forward.

f. Airfields and Air Facilities

The MAGTF ACE operates from either existing airfields or air facilities within or close to the MAGTF objective. The ACE fixed-wing aircraft may require runway surfaces as long as 10,000 feet.

Fixed-wing aircraft can operate from runways as short as 4,000 feet by reducing fuel and ordnance loads and by using arresting gear. Helicopter, short takeoff, vertical landing, and tiltrotor aircraft runway requirements are considerably less. Less-developed strips can be enhanced with expeditionary airfield (EAF) equipment. If required, and if time permits, a complete EAF can be installed.

An EAF is a prefabricated and portable airfield. The effort (e.g., materiel, engineer support, operational guidance, security) needed for the installation and operation of an EAF may require the support of all MAGTF elements. When deployed, EAFs provide the capability to launch and recover MAGTF aircraft under all weather conditions. Expansion of EAF facilities into a strategic expeditionary landing field allows the support and maintenance for a complete wing-sized ACE. The strategic expeditionary landing field has parking and taxiways to accommodate the Air Mobility Command (AMC) and Civil Reserve Air Fleet aircraft.

Bare base EAFs provide the capability for using an existing airfield or road network to establish an EAF. Bare base EAFs are established in place of a full EAF due to extensive embarkation or construction requirements associated with the full EAF and due to the speed in which a bare base air facility can be made operational in support of MAGTF sorties. The bare base EAF concept calls for the use of available concrete and/or asphalt-surfaced facilities and involves embarking only those assets necessary to conduct air operations; e.g., airfield lighting or marking, landing aids, or arresting gear. Bare base kits have been established to support all EAFs.

Small EAFs are constructed by the MWSS. Larger airfields may require the MWSS to be augmented by the LCE ESB or NCFs. The NMCB provides augmentation to the MAGTF's general engineering capability, or, if required, assumes full responsibility for construction of the EAF. Each maritime prepositioning ships squadron (MPSRON) contains EAF equipment to provide a capability of airfield lighting, expeditionary arresting gear, and airfield landing matting. For additional information on EAF capabilities, see MCWP 3-21.1.

g. Seaports

Seaports of embarkation (SPOEs)/seaports of debarkation (SPODs) are critical transportation installations within the tactical-level logistic effort. They serve to connect the MAGTF to external sources of supply (greater supply chain management and distribution) and support the force's deployment/redeployment. The availability of ports suitable for amphibious ships, commercial ships, and maritime prepositioning ships (MPSs) is another key planning consideration. Ports with well-developed infrastructure can significantly increase the buildup and throughput of MAGTF personnel, supplies, and equipment ashore. When sustained operations ashore are anticipated, ports often become a CSSA.

Chapter 2
Logistic Functional Area Support Operations

This chapter provides an overview for each of the tactical-level logistic functional areas within CSS. To support tactical-level operations, logisticians commonly discuss support requirements in terms of functional areas, processes, and plans for each area. Although logisticians develop separate systems and plans for each functional area, all functions must be integrated into the overall logistic support effort.

Section I. Supply

Supply consists of procurement, requisitioning, distribution, and maintenance while in storage, and salvage of supplies, including the determination of kind and quantity of supplies. Providing materials to equip, support, and maintain a military force is part of the supply cycle. This section addresses the various supply classes and subfunctions available to support tactical-level operations.

2101. Logistics Combat Element Supply Support Operations

The LCE commander's primary concern is providing the MAGTF commander with a supply capability and resupply when required.

a. Landing Force Supplies
Landing force supplies are the supplies and equipment in the assault echelon and the assault follow-on echelon (AFOE) of the amphibious task force (ATF). They sustain the landing force until a distribution pipeline is established from the supporting establishment to the theater of operations. Predeployment planning determines the type and quantity of landing force supplies. The categories of landing force supplies are the basic load, prepositioned emergency supplies, and remaining supplies.

(1) Basic Load. A basic load consists of the types and quantities of supplies that assault forces carry to a specific mission, including the supplies carried by individuals. Usually, basic loads are expressed either as days of supply or days of allowance. The basic load may change as the tactical situation dictates. There may be a basic load for landing and a different basic load for operations ashore. The basic loads for surface and assault support forces may be different. The basic load should not exceed the capabilities of a unit's organic transportation or the commander's estimate of supply requirements for combat.

(2) Prepositioned Emergency Supplies. The commander prepositions supplies on ships for emergent replenishment during ship-to-shore (STS) movement. These on-call supplies are available for immediate delivery to units ashore via surface or aerial delivery. Emergent

replenishment is categorized as either floating dumps or pre-staged vertical takeoff and landing (VTOL)-lifted supplies.

The commander pre-stages VTOL-lifted supplies to support units that were delivered via vertical assault, and, if required, the supplies can be used to support surface assault units. Pre-staged VTOL-lifted supplies are pre-packaged, high-priority supplies. Like floating staging areas, these supplies are available on-call for units ashore. Requests for this category of supplies are made by the unit to the tactical-logistical (TACLOG) group. After the initial stages of the assault, remaining supplies are used to expand supply staging areas ashore. Both pre-staged VTOL-lifted supplies and floating staging areas may be assigned landing serial numbers to help identify and deliver specific materiel. See *MCWP 4-11.7* for further discussion.

(3) Remaining Supplies. Supplies that are not part of the basic load or prepositioned emergency supplies are used in general support to the MAGTF. They constitute the major portion of the supplies transported to the operational area in the assault echelon and the AFOE. When transitioning from operational maneuver from the sea to sustained operations ashore, the commander uses these supplies to build stockage ashore. The LCE unloads the bulk of remaining supplies during general unloading using a combination of surface and vertical lift platforms.

b. Sustainment

Sustainment involves additional supplies provided to the LF from outside agencies. Sustainment sources include joint, interagency, intergovernmental, and multinational support.

c. Ground Supply Operations During the Amphibious Assault

(1) Landing Force Support Party. The landing force support party (LFSP) mission is to perform specified combat service support operations for the landing force during the amphibious assault. It also provides initial landing support and combat service support to the landing force during the amphibious operation. It is a temporary task organization composed of Navy and Marine Corps elements tasked to provide initial combat service support during the STS movement. The LFSP's strength and composition are determined during the amphibious operation's initial planning phase. The LFSP may include units or detachments from the GCE, ACE, LCE, and the Navy. The MAGTF's organization and mission, the number of landing beaches/zones through which the MAGTF will land, and the mission and size of the units assigned to the beaches/zones will determine the LFSP's configuration. The LFSP is under the operational control (OPCON) of the commander, landing forces.

(2) Tactical-Logistical Group. The TACLOG group is a temporary LF organization established at each level of the Navy STS control organization. The TACLOG group advises the Navy control groups of LF requirements for supporting the surface and vertical assault forces during STS movement. The TACLOG group monitors STS movement and helps the Navy control the movements of scheduled waves, on-call waves, and nonscheduled serials.

(3) Supply During the Assault. Initial assault units will request supplies directly from the TACLOG group until a shore party or HST is established ashore. At that point, assault units submit CSS requests for supplies to either the shore party or HSTs. The teams either fill or relay requests to the TACLOG group. In an emergency or when communications fail, the assault element may pass requests directly to the TACLOG group. See figure 2-1.

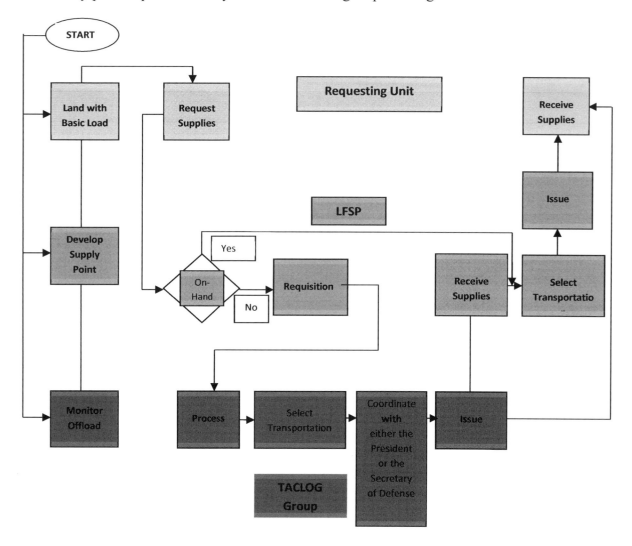

Figure 2-1. Supply Support During the Amphibious Landing.

(4) Shore Party Supply Operations. After the shore party group lands, it establishes inland supply staging sites. It controls the receipt of selective unloading. Shore party group and HST supply personnel unload, sort, store, safeguard, and issue supplies. Shore party teams and HSTs distribute supplies directly to the consumer by using the fastest available means.

(5) Critical Items. If a critical item is not on-hand, the shore party or HST notifies the TACLOG group. The TACLOG group then locates the item and coordinates transportation from the Navy control organization.

(6) Prioritization. Before the Navy assigns transportation to move unscheduled supplies ashore, the TACLOG group must determine the impact on the tactical situation. It must assess the priority against the priority for landing scheduled and on-call serials.

(7) Helicopter and VTOL Delivery. The shore party or HST receives supplies and distributes them to the requestor. Delivery can be directly from the ship by aircraft to the end user.

d. Ground Supply Operations During Subsequent Operations Ashore

Battalions and aircraft groups have organic supply capability. Marine Corps and/or Navy directives and local operating procedures dictate the procedures that units with organic supply capabilities use to request replenishment in combat. Figure 2-2 depicts management and execution of ground supply operations during subsequent operations ashore.

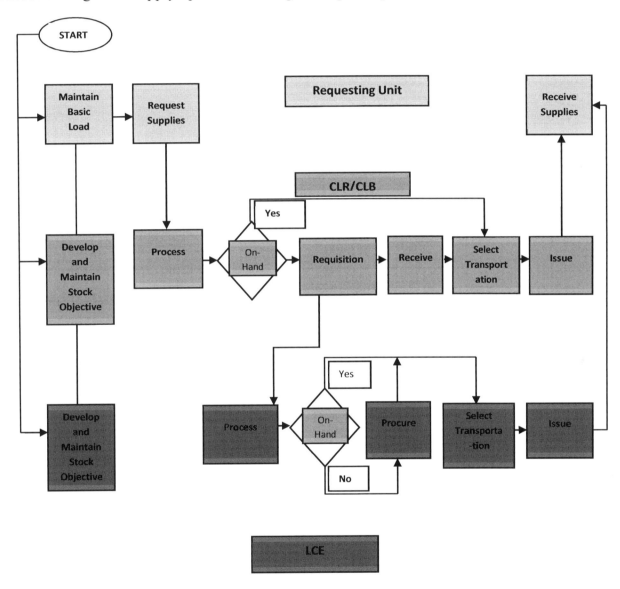

Figure 2-2. Supply Support During Subsequent Operations Ashore.

(1) User Request Support. Locally established manual procedures are the norm for initial request from consumers. On receipt of user request, the supporting LCE intermediate supply activity determines whether the item is on hand. If item is available, the LCE transports it to users via unit distribution. Consumers on supply point distribution are notified where and when they can pick up the item. If the item is not on hand, the LCE passes the requisition to the next higher level. The LCE will keep the requesting unit informed about the status of the pending requisition until the item is distributed to the consumer.

(2) LCE Support. The LCE receives requisitions from subordinate LCE or directly from the user. The LCE uses formal procedures for both stock replenishment and passing unfilled user requests to other logistic support organizations. Where possible, LCEs use automated systems to pass and track both requisitions and reports. During the early stages of an operation before automated systems are established, the LCE uses manual requisition procedures.

(3) Unfilled Requisitions Support. The LCE in theater passes unfilled requisitions to an in-theater source, if available, or to the MLG or Marine Corps supporting establishment in CONUS. Marine Corps user manuals and MAGTF OPORDs establish specific supply procedures for LCEs during operations.

(4) Mode of Transportation. The LCE normally selects and provides the mode of transportation to deliver supplies and equipment to supported and supporting units. Although the LCE selects the mode of transportation, the consumer may influence the decision by providing viable information to the LCE to accommodate a more suitable mode of transport based on priority and available resources. For example, a request for a rapid ammunition resupply from a unit preparing to repel an imminent attack would probably justify the use of airlift transport.

(5) Delivery Method. Direct shipment to the consumer is the best method of delivery. Bypassing intermediate CSS installations reduces the number of times that equipment or supplies are handled (reducing the potential of shipping damage) and achieves more-responsive delivery to the supported unit. Sometimes, direct shipment is not an option, and supplies must be delivered to the supporting combat logistics battalion (CLB). This method achieves transportation economies when moving large bulk quantities by taking advantage of opportunity lift. Rather than hauling a partial load, trucks can carry noncritical supplies to the LCE for later movement to the consumer.

(6) Distribution Method. The MAGTF commander usually determines the distribution method used, and the method is reflected in annex D of the OPORD. When supply point distribution is used, care must be taken not to restrict operations for units that have limited organic transportation. When the MAGTF commander selects unit distribution, the LCE/ACE commander must develop a transportation network from the supported organization to the rear supply area that does not generate equipment shortages in rear areas. As a general rule, the MAGTF commander must use a combination of supply point and unit distribution. Highest priority for unit distribution is usually given to engaged units having limited organic transportation. Engaged units having organic transportation are usually the next priority. Units not in contact with the enemy usually receive the lowest priority. Distribution methods will be covered in greater detail in chapter 3.

2102. GROUND COMBAT ELEMENT SUPPLY SUPPORT OPERATIONS

Figure 2-3 depicts a tactical situation in which an LCE is in direct support of GCE units. The LCE establishes liaison with the infantry regiment. Requests from the battalions go directly to the LCE, which issues supplies based on the supported commander's priorities and allocations.

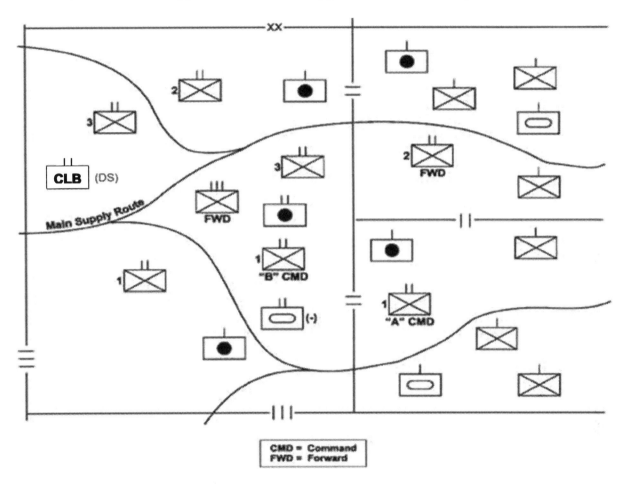

Figure 2-3. Notional Supply Support During Subsequent Operations Ashore.

a. Commander's Flexibility

The supported commander can organize his unit in a variety of ways to accomplish their mission. For example, the commander may divide the headquarters into A and B command groups and/or position the organic logistics differently than previously described. The commander should position organic logistics forward of the supporting CSS installation. The ground unit supply train is a means of internally task-organizing and employing the logistic assets of tactical units.

When employing combat trains, some of the GCE unit's organic logistic capabilities are forward. Maintenance contact team repairmen, ammunition technicians, and supply personnel are with the supply trains to provide front-line support. Routinely, the unit establishes a main command post with essential elements that support tactical operations. The commander locates most of the unit's logistic capability with the unit or field train. Often the commander locates these trains with the supporting LCE.

Finally, all units have administrative elements located behind the GCE rear boundary. In the administrative rear, supply and warehousing personnel distribute individual equipment and care for tentage, personal effects, and other equipment not required for the sustainment of combat operations. Table 2-1 shows a generic example of a typical battalion in combat.

Table 2-1. Battalion Task Organization for Combat.

Rear	Main Command Post	Forward Command Post
S-1/adjutant	Executive officer	Commanding officer
Supply chief	Headquarters commandant	S-2
Administrative supply clerks	S-4A/S-4 chief	S-3
Casualty replacements	Motor transport officer	Fire support coordinator
	Ordnance officer	S-4
	Supply officer	Communications officer
	Organic logistics	Organic logistics

b. Supply Trains

Supply trains serve as the link between forward tactical elements and the supporting LCE. The use of trains enables logistics to be performed as far forward as the tactical situations permit. Depending on the situation, trains may provide logistics to the battalion's organic and attached units. Trains may be fully mobile; however, trains are usually movable rather than mobile. In the Marine Corps, this concept applies to unit, battalion, and regimental trains.

(1) Unit Trains. Unit trains centralize the unit's organic logistic assets. These trains are most appropriate in defensive, slow-moving, or static situations. The commander uses this option when a tactical situation dictates self-contained train operations for centralization and control. For example, during the early phases of an amphibious operation, the battalion must locate its logistic capability in the BSA or landing zone. The use of unit trains in this situation provides simplicity, economy, and survivability against ground attack.

(2) Battalion Trains. Normally, to improve responsiveness, flexibility, and survivability against air attack, trains supporting battalion-sized units are echeloned into combat trains and field trains.

(a) Combat Trains. Combat trains are organic elements that provide critical logistics in forward areas. Mobility is the key for combat trains, which are kept as small as possible to move with the supported forces. A combat train's survivability depends on its small size and its own firepower. Usually, a combat train—

- Transports some battalion supplies with limited medical supplies.
- Carries MCTs.
- Hauls rations, fuel, ammunition, and critical spare parts.

(b) Field Trains. Field trains consist of the battalion's remaining logistic assets and are located closer to the rear than the combat trains. Field trains may carry the battalion aid station, the mess section, and the supply section.

(3) Regimental Train. The regimental train consists of the logistic assets required to sustain the regimental headquarters and attached units under the direct control of the regiment. Logistics needed by combat units should be allocated to battalion trains, and logistics that are not time-critical can be consolidated in the regimental train.

(4) Positioning Considerations. Logistic principles of responsiveness and survivability should be the main considerations when selecting a train site. In general, trains should be located—

- On defensible terrain to allow the best use of limited personnel assets.
- In an area with enough space to permit dispersion.
- In an area that provides concealment.
- On firm ground to support heavy vehicle traffic.
- Near a suitable landing zone.
- Close to main supply routes.
- In an area that allows good communication.

(5) Positioning Responsibility. The S-4 coordinates with the executive officer, headquarters commandant, and S-3 in selecting train locations. When the train collocates with another element, such as the supporting LCE, the S-4 must also coordinate with that element. This option improves coordination and security. Turnaround time, communications requirements, or other mission-related considerations may necessitate locating the trains elsewhere.

(6) Train Displacement. Proper positioning of trains minimizes displacements and increases the quantity and quality of support. When displacing trains, the S-4 selects the technique that best complements the battalion's tactical operations. Trains may be displaced concurrently with the displacement of the tactical elements or by echelon. Echeloned displacement enhances continuity of logistic support.

(7) LCE Trains. Trains are employed in numerous ways by LCE units in the resupply process. Figure 2-4, on page 2-9, illustrates train techniques that are commonly used during resupply operations. The distances provided in figure 2-4 would be reduced for close terrain (i.e., urban or jungle) or expanded for high enemy threat. The LCEs trains may move forward to resupply unit trains, which resupply the using units. The LCEs trains are positioned where most responsive, yet survivable.

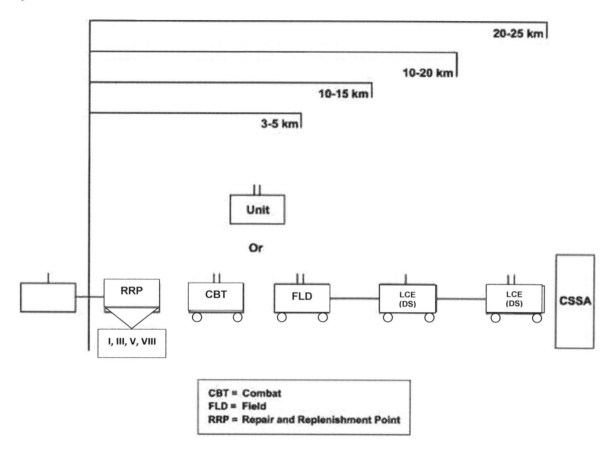

Figure 2-4. Train Techniques Commonly Used During Supply Operations.

(8) Replenishment Methods. The service station and tailgate issue methods are the two most common methods used to replenish unit trains.

(a) Service Station. The service station method in figure 2-5, on page 2-10, involves vehicles leaving their tactical positions and entering an established resupply area. The number of vehicles being resupplied at one time depends on the enemy situation and resupply capabilities. The resupply area is designated as a series of resupply points for vehicles. Traffic flow through the resupply area is one way to enhance efficiency. After completing resupply, the vehicles move to the holding area for a pre-combat inspection, if time permits.

MCTP 3-40B. Tactical-Level Logistics

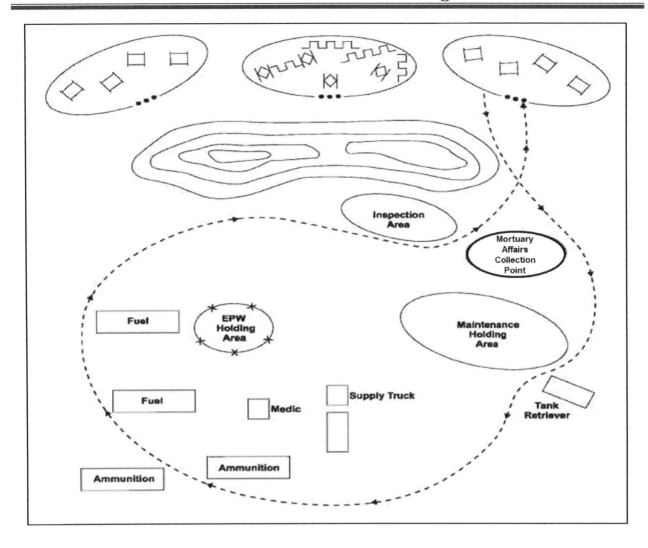

Figure 2-5. Service Station.

(b) Tailgate Issue. The tailgate issue method is normally conducted in an assembly area. This method involves resupply while combatants remain in their positions. Vehicles stocked with petroleum, oils, lubricants, and ammunition stop at each individual vehicle position to conduct resupply services. This method places the resupply vehicles at greater risk, but maintains tactical positioning and reduces traffic flow. If the tailgate issue method is used in forward positions, then resupply must be masked by the terrain. See figure 2-6 on page 2-11.

MCTP 3-40B. Tactical-Level Logistics

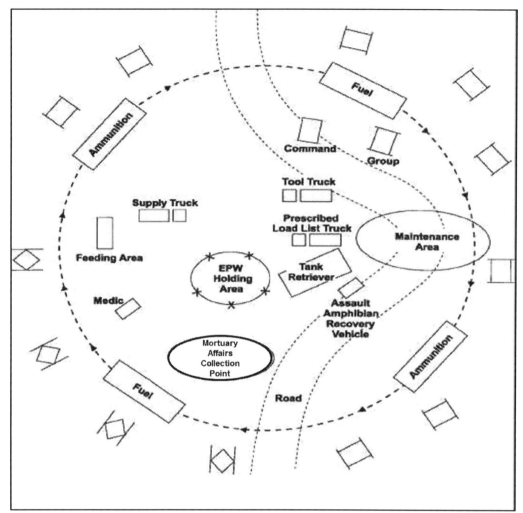

Figure 2-6. Tailgate Issue.

2103. AVIATION-PECULIAR SUPPLY SUPPORT OPERATIONS

The Navy provides supply support for all Navy-funded aircraft and aviation support equipment in the ACE. The Marine Corps supply system provides ground supply support to aviation elements. For aircraft ammunition, the source of supply is either the Navy or a theater activity. The LCE distributes aircraft fuel to the MWSS operating the fuel dispensing system at an air FARP or air facility. The LCE distributes Class V (A) to the MALS, which operates the aviation ASP at an air facility/FARP.

a. Marine Aviation Logistics Squadron

When a MAG deploys, the MALS is the focal point for aviation supply and maintenance. The MALS supply and maintenance departments manage aircraft consumable and reparable parts and supplies. The MALS supply department receives and processes requisitions for all units. If the item is not in stock, the MALS passes the requisition to the naval supply activity (inventory control point) in the theater support area, which either fills the request or forwards it to the appropriate source in CONUS or to an adjacent theater's naval supply activity.

Elements of the MALS can be embarked aboard ship as part of an existing naval aircraft intermediate maintenance department. It can also be embarked aboard a special-purpose strategic sealift vessel, the T-AVB. The T-AVB can be employed in one of three operational configurations depending on the operational needs of the ACE commander. For further detail, see MCWP 3-21.2, *Aviation Logistics*.

b. Replacement Aircraft

The squadron requests replacement aircraft and depot-level repair of aircraft. It passes the request for replacement aircraft to the aircraft group, which passes it to the MAW. The ACE passes the request to the type commanders (MARFORCOM and MARFORPAC ALD and/or Commander, Naval Air Forces). The MALS, MAG, MAW, and type commanders coordinate placement of aircraft into depot maintenance. The transferring activity is responsible for flying replacement aircraft directly to the receiving squadron or to an airfield near the receiving squadron. The receiving squadron accepts the aircraft and reports the aircraft's status to the MAW.

c. Aircraft Fuel and Ammunition

The LCE normally establishes a bulk fuel area ashore from which it draws fuel to deliver to the MWSS which, in turn, dispenses fuel to aircraft. Similarly, the LCE normally establishes one or more centralized ASPs for the purpose of receiving, accounting, storing, and issuing Class V materiel. Centralized ASPs are generally supported by ammunition technicians provided by the LCE, along with a small cadre of aviation ordnance technicians, who assist in the throughput of Class V (A) to outlying satellite ACE ASPs. Figure 2-7, on page 2-13, shows the relationship between aviation units and the LCE for ground supply support and for aircraft fuel and ammunition support. Satellite ASPs are generally established for both air and ground units in an effort to minimize the effects of time and distance on the efficient delivery of munitions to the end user.

Satellite ASPs used to support the ACE should be collocated at the ACE's operational air facility (facilities). The ACE (MALS) aviation ordnance department is staffed for and fully capable of all functions similar to those performed by a centralized ASP. The MALS aviation ordnance department is responsible for establishing, operating, and maintaining satellite ASPs for Class V (A). All Class V (A) materiel arriving at the airfield is received and stored under the direction of the ACE aviation ordnance department unless accompanying documentation specifies further transportation to either a centralized ASP or another satellite ASP.

Aviation ordnance personnel can augment LCE ammunition company on a contingency basis. The augments from aviation ordnance department within MALS should be knowledgeable of aviation ordnance peculiarities and different inventory reporting requirements that exist for Navy-owned ammunition. These personnel are assigned to the ASP nearest the SPODs responsible for storing and distributing Class V (A) and (W) arriving in theater. They assist in the receipt, segregation, storage, and distribution of Class V (A) within the theater of operations.

MCTP 3-40B. Tactical-Level Logistics

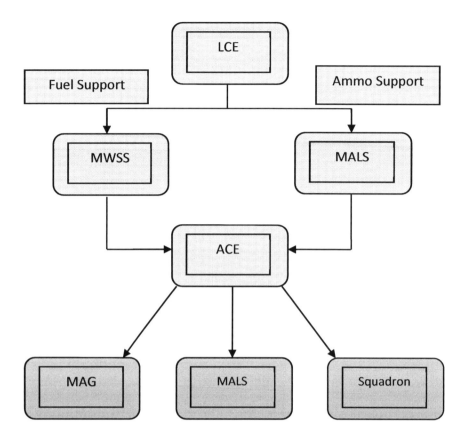

Figure 2-7. Ground Supply Operations for Aircraft Fuel and Ammunition Support.

SECTION II. MAINTENANCE

Maintenance involves those actions taken to repair or restore equipment to serviceable condition. While the purpose and functions of equipment maintenance are universal, the Marine Corps has developed applications for the support of ground-common and aviation-peculiar equipment. This section describes maintenance support for the levels, categories, and subfunctions described in chapter 1.

2201. GROUND MAINTENANCE SUPPORT OPERATIONS

a. Levels of Maintenance
Marine Corps maintenance capability is defined within two levels of maintenance: field and depot. The distinction between field and depot levels of maintenance is based on the maintenance tasks performed within each.

(1) Field Levels of Maintenance. Field maintenance is any maintenance that does not require depot maintenance capability and is performed by crew/operators and maintainers within

Marine Corps organizations and activities, and/or by approved commercial/contract sources. Maintenance tasks performed within the field level of maintenance are categorized as organizational and intermediate. Alignment of tasks within the field level of maintenance is based on supporting/supported relationships and respective capabilities among units. A unit may perform any field maintenance tasks for which it is manned, trained, and equipped. Units are not authorized to conduct maintenance tasks outside of their assigned capabilities.

(a) Organizational Maintenance. Organizational maintenance tasks are the responsibility of and performed by a using organization on its assigned equipment. They normally consist of inspecting, servicing, lubricating, and adjusting, as well as replacing parts, minor assemblies, and subassemblies. Both maintenance personnel and equipment operators accomplish this. The MCT conducts recovery, evacuation, and repair. They determine whether an item is reparable at the recovery site. The MCT either fixes the item, requests parts and an intermediate-level MST from the LCE, or supervises the item evacuation.

(b) Intermediate Maintenance. Intermediate maintenance tasks may require a higher level of technical training, specialized tools, and/or facilities. They consist of a range of capabilities, including modification, replacement, fabrication, repair, and calibration of equipment or items. The MST is an intermediate maintenance version of the MCT. The MST inspects, diagnoses, classifies, and repairs equipment at forward sites.

(2) Depot Levels of Maintenance. Depot maintenance actions are those taken on materiel or software involving the inspection, repair, overhaul, or modification or reclamation (as necessary) of weapons systems, parts, components, assemblies, and subassemblies that are beyond field maintenance capabilities, and/or that are authorized and directed by DC I&L. Depot maintenance is not defined by location. The Marine Corps' organic depots, other service depots, commercial industrial facilities, original equipment manufacturer, or a combination thereof, may perform depot maintenance-related activities throughout the logistic chain framework.

b. Maintenance During the Amphibious Assault

Assault force elements land with a few organizational maintenance personnel. The majority of the organizational maintenance capability lands in nonscheduled waves. Once the first assault waves are ashore, the LFSP provides the only significant maintenance capability.

Although the LFSP has limited recovery, evacuation, and repair capabilities, it has a small block of critical repair parts tailored to match the quantity and type of equipment in the assault waves. The LFSP replaces components and assemblies rather than repairing them. It uses selective interchange to offset the limited depth and breadth of repair parts. One of the first tasks of the LFSP maintenance detachment is to establish maintenance and salvage collection points.

The LFSP must develop an aggressive recovery and evacuation plan because extensively damaged items may provide repair parts for other essential items. Damaged equipment should be placed on resupply vehicles returning to the LFSP. Assault elements should abandon equipment only when the tactical situation prevents recovery. When unable to recover equipment, units should report the location of the item to the LFSP for later recovery and evacuation.

c. Maintenance During Transition Periods

When the tactical situation ashore stabilizes, the MAGTF commander lands nonscheduled units such as unit trains with the organizational maintenance elements. As the assault units' organizational maintenance capability expands, the LFSP shifts its efforts to intermediate maintenance. Assault units normally position their trains near the LFSP to permit mutual support, to avoid duplication of facilities, and to reduce the transportation burden.

d. Maintenance During Subsequent Operations

When appropriate, the MAGTF commander lands the additional LCE units. Once the LCE is established ashore, the MAGTF commander disestablishes the LFSP. After the AFOE arrives, the LCE commander reaches full maintenance capability. When the maintenance unit cannot repair an item, it evacuates the item to the next higher level.

e. Organizational Maintenance

Units owning equipment have organizational maintenance responsibilities. Proper maintenance is essential to sustain combat operations. The MCT is the centerpiece of organizational maintenance.

(1) Maintenance Contact Team Capabilities. The MCT consists of organizational maintenance repairmen with tools, test equipment, and critical, high-usage repair parts. These repairmen inspect, diagnose, classify, and repair equipment at forward sites. In addition, the MCT may include communications, engineer, motor transport, or ordnance repair personnel. The logistic officer determines the exact number of Marines and mixes of skills in MCTs and positions them in the appropriate train. When using combat trains, MCTs are forward where they are more responsive to the tactical unit. If deployed with a unit train, MCTs are farther to the rear.

(2) Maintenance Contact Team Operations. The MCT conducts recovery, evacuation, and repair. They determine whether an item is reparable at the recovery site. The MCT either fixes the item, requests parts and an intermediate-level MST from the LCE, or supervises the item evacuation. Figure 2-8, on page 2-16, shows relationships between various maintenance agencies. The GCE collection points represent the battalion and regimental trains.

f. Intermediate Maintenance

The three elements of an intermediate maintenance concept are the MST, the LCE forward maintenance detachment, and the MLG intermediate maintenance activity (IMA).

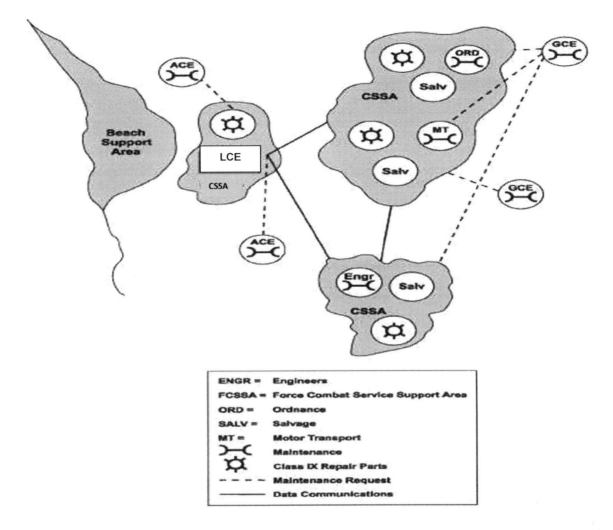

Figure 2-8. Ground Equipment Maintenance Process in Combat.

(1) Maintenance Support Team. The MST is an intermediate maintenance version of the MCT. The MST has intermediate maintenance repairmen with tools, test equipment, repair parts, and likely a wrecker or maintenance vehicle. These repairmen inspect, diagnose, classify, and repair equipment at forward sites. The LCE operations officer determines the number of Marines and mix of skills per team. Normally, MSTs move forward to repair a specific item of equipment. This technique allows the MST to draw the needed parts and tools before moving based on input from the MCT.

(2) Logistics Combat Element Forward Maintenance Detachment. The LCE forward maintenance detachment is the element of an LCE that operates the maintenance facilities and collection points far forward. The forward support maintenance detachment—

- Evacuates inoperable equipment from supported units' collection points.
- Performs intermediate maintenance within its capabilities.
- Provides repairman, tools, and test equipment to MSTs.

(3) Marine Logistics Group Intermediate Maintenance Activity. The MLG intermediate maintenance activity (IMA) provides robust principal end item repair and component rebuild support to the MAGTF. The MLG commander establishes a centralized IMA in a CSSA to perform complex, time-consuming maintenance activities during sustained operations ashore. The LCE commander forms multiple on-call MSTs and, during surge periods, sends them forward either to assist MCTs or to augment the LCE forward maintenance detachments.

g. Recovery, Evacuation, and Repair Cycle

These capabilities differ during the various phases of combat operations and increase as more of the MAGTF lands. Thus, standard maintenance processes must be established to reestablish operable equipment.

(1) Recovery Responsibility. As much as capability and the tactical situation allows, the owning units are responsible for retrieving immobile, inoperative, or abandoned materiel. They move recovered equipment to a maintenance collection point or a main supply route.

(2) Evacuation. If neither the owning unit nor the LCE can repair a recovered item, the LCE evacuates it. If the MAGTF commander authorizes selective interchange, the LCE may remove and use parts before evacuating an item. The LCE evacuates recovered equipment directly to a designated repair or disposal agency.

(3) Recovery Considerations. Commanders should closely monitor and control recovery and evacuation operations. Logistic officers must establish recovery and evacuation priorities and carefully allocate personnel and equipment to these operations. For example, combat vehicles, weapons, and weapons' platforms often have a higher recovery priority than other items. Also, the extent of damage affects recovery priority. When the unit must recover two or more of the same item, the item requiring the least repairs should be recovered first. If material is in danger of capture, the owning unit should recover all salvageable parts and components and destroy the remaining equipment. The following is a suggested recovery priorities list:

- Items immobilized by terrain.
- Items with failed or damaged components that require little repair.
- Damaged items that require significant expenditure of recovery and repair effort to return them to operation.
- Contaminated items that require significant recovery, repair, and decontamination effort
- Salvageable items.
- Enemy material.

(4) Positioning. Combat and combat support unit commanders should position their recovery capability forward. As a rule, the recovery capability consists of personnel and equipment organized in MCTs. The LCE commanders distribute maintenance assets to achieve a balance between economy and responsiveness.

2202. AVIATION-PECULIAR MAINTENANCE SUPPORT OPERATIONS

The MALSP and the MPF program (including aviation logistic support ships) provide aircraft support personnel with the ability to sustain all aircraft types that comprise a MAGTF ACE. Specifically, these programs enable aviation logisticians to identify and integrate the people, aircraft support equipment, mobile facilities and/or shelters, as well as spares and/or repair parts needed to support a MAGTF ACE.

a. Marine Aviation Logistics Support Program

Most Navy funded logistical support for aviation units is provided under MALSP. The primary objective of MALSP is to expedite the delivery of required aviation-peculiar logistics to support any contingency. The MALSP and MPF provide a building-block method of quickly task-organizing, deploying, and sustaining ACE aviation-peculiar assets by structuring aviation logistical support into contingency packages that can be phased into an operating area.

(1) Support Packages. The MALSP provides comprehensive and replenishable sustainment packages while reducing lift requirements and force closure time. These support packages are used as building blocks to keep aircraft operational during every phase of an operation. For further discussion, see MCWP 3-21.2.

(a) Fly-In. Fly-in support packages (FISPs) can be viewed as enabling packages. They provide the organizational-level spare parts support that allows Marine aircraft to commence flight operations immediately on arrival in theater. The FISPs are airlifted to the operating site as part of the FIE. They are combined with the organizational-level and/or limited intermediate-level aircraft support equipment transported aboard MPF ships. This combination of assets is capable of providing critical aviation support for 30 days of combat flying. If flight operations require more than 30 days of spare parts support, then contingency support packages (CSPs) are provided to augment the FISP.

(b) Contingency. The CSPs augment the FISPs by adding common maintenance support items which are used by more than one Marine aviation unit and peculiar maintenance support items used for a specific aircraft or support equipment application. These packages support both organizational- and intermediate-level maintenance. The CSPs integrate the maintenance equipment, mobile facilities, spare parts, and personnel to support and sustain each type of deployed tactical Marine aircraft. Rapidly deployable organizational-level individual material, mobile facilities allowances, and personnel allocations are identified in master allowance documents for each aviation element. The master allowance documents consist of TOs, individual material readiness lists (IMRLs), tables of basic allowance (TBAs), aviation consolidated allowance list (AVCAL), and coordinated shipboard allowance list (COSAL). The CSP allowances are computed at the combat flying-hour rate for a 90-day endurance period and

are supplemental allowances to those identified in AVCAL, COSAL, IMRL, and TBA. The CSP allowances, which are derived from the master allowance documents, are separated into the following subcategories:

- **Common Allowances.** Common CSP allowances consist of those Marine common assets that the rotary- or fixed-wing MALS of an ACE provide to support the majority of assigned aircraft. A fixed-wing Marine common item is one that has application to at least the F/A-18 and AV-8B aircraft, which are part of an ACE. A rotary-wing common item is one that has application to at least the CH-53E, MV-22, and AH-1W aircraft, which are a part of an ACE. Weight, cube, cost, reliability, and supportability are the primary considerations in determining what parts are included in the CSP. For planning purposes, it is assumed that the fixed- and rotary-wing MALS will be geographically separated.
- **Peculiar Allowances.** Peculiar CSP allowances consist of those maintenance items required for intermediate-level support of a specific type/model/series (T/M/S) aircraft and of associated support equipment that a MAG provides to a MAGTF ACE.

(c) Follow-On. Follow-on support packages represent the final MALSP building block. The introduction of the follow-on support package would, in essence, provide ACE aircraft with the same support received in garrison.

(2) Reconfiguration for Deployment Support. Marine aircraft squadrons of a particular T/M/S aircraft are generally consolidated and attached to only two or three MAGs. To form an ACE, one or more fixed-wing and tiltrotor/rotary-wing MAGs reconfigure into a task-organized fighting unit by retaining or attaching only mission-essential aircraft, aircrew, and operations support personnel and equipment. Under MALSP, aviation logisticians identify people, IMRL items, TBA, and AVCAL and/or COSAL allowances that are needed to support the quantities of each T/M/S of aircraft being detached and attached to ensure that reconfigured MAGs include the necessary MALSP resources.

(3) Support Personnel Requirements. Staffing and organization are two personnel considerations in support of the MALSP.

(a) Staffing. Without adequate staffing of qualified maintenance, supply, and administrative personnel, this program would not succeed. The MALS and supported squadrons' TOs should provide the right quantity of skilled personnel to support a task-organized ACE.

(b) Organization. Each MALS is organized to provide a core intermediate-level capability of supervisory and common support personnel necessary to maintain fixed-wing or tiltrotor/rotary-wing aircraft that join an ACE. The MALS T/O contains the personnel component of a common CSP, which forms the nucleus of an ACE allowance list (fixed-wing or tiltrotor/rotary-wing). Each tactical aircraft squadron's T/O has a separate listing of intermediate-level billets that consist of military occupational specialty (MOS) skills that are peculiar to that squadron's T/M/S aircraft. The MALS provides the MAGTF commander with the capability to support the peculiar requirements of the T/M/S aircraft assigned to that ACE. Whenever the MAG detaches aircraft and sends them to an ACE, a unit deployment, or an exercise, the MALS uses the intermediate

maintenance portion of aircraft squadron TOs and produces a complete CSP (i.e., IMRL, AVCAL, COSAL, TBA) for the receiving MALS.

b. Aviation Logistics Support Ship

The T-AVB concept was developed to transport critical intermediate-level maintenance and supply assets to a forward operating area in support of deployed Marine aircraft. The primary mission of the T-AVB is to provide dedicated sealift for movement of intermediate-level logistic support for use in the rapid deployment of a MAGTF ACE. A secondary mission, to serve as a national asset dedicated to strategic sealift, can be exercised if the embarked MALS is phased ashore. To enhance responsiveness, one ship is berthed on the East Coast and another on the West Coast of the United States. Both ships can be configured to allow for tailored intermediate-level repair capability while underway, in stream, or pier-side. For further discussion, see MCWP 3-21.2.

c. Maritime Prepositioning Ships

The MPF program provides fleet commanders with deployment flexibility by including field level of maintenance and limited intermediate-level aviation support equipment and Class V (A) in each MPF squadron.

Maritime prepositioning ships are roll-on and roll-off, civilian-crewed, Military Sealift Command-chartered ships that are organized into two MPSRONs. In peacetime operations, they are usually forward deployed in strategic locations worldwide. Currently, MPSRON-2 and MPSRON-3 are composed of seven ships respectively.

(1) Capabilities. Each MPSRON has a fixed set of embarked equipment and supplies. Generally, this set contains sufficient quantities of supplies (except Classes VI and X) to sustain a MEB for 30 days of combat operations. To support ACE operations, each MPSRON contains a tailored set of organizational-level aircraft support equipment for each T/M/S aircraft assigned to the supported ACE. Additionally, each MPSRON includes limited intermediate-level facilities equipment. This equipment is designed to provide common intermediate-level functions normally associated with the MALS (e.g., tire and wheel buildup, battery maintenance). On arrival at the port of debarkation, aircraft equipment will be off-loaded, and when combined with the equipment embarked aboard the FIE, T/M/S aircraft FISP allowances, and support personnel, the ACE will be capable of sustained combat flight operations for up to 30 days or, if augmented, until the arrival of the host MALS via the T-AVB.

(2) Unique Features. The association of specific forces with their prepositioned materiel is a unique feature that sets apart the MPF program from other afloat prepositioned programs. This critical association facilitates the rapid employment of materiel in support of expeditionary operations. The strategic stationing of MPSRONs contributes to worldwide responsiveness and provides the ability to mass a large force at one point by using several squadrons and associated forces.

SECTION III. TRANSPORTATION

Transportation is movement from one location to another by using highways, railroads, waterways, pipelines, oceans, and air. Transportation is needed to deliver initial sustainment materiel and personnel in the correct locations at the proper times to start and maintain operations. Any major disruption of transportation support can adversely affect a MAGTF's capability to support and execute the assigned mission.

2301. MOTOR TRANSPORT OPERATIONS

Motor transport operations may be either combat support or CSS. The commander may attach motor transport units to supported units. The commander may also control allocated motor transport resources by assigning an appropriate mission. Successful motor transport operations require careful management.

Economical transportation operations dictate matching the number and type of vehicles to the task and reducing the turnaround time. Factors that affect turnaround time are distance, rate of march, and the time it takes to load and unload. The turnaround time can be delayed if shippers and receivers responsible for loading and unloading vehicles are slow or fail to release the vehicles after unloading. For further discussion, see MCWP 4-11.3.

a. Operational Techniques

The commander may increase the tonnage moved with a fixed number of trucks by adopting some or all of the following techniques:

- Loading each vehicle to its maximum allowable capacity.
- Increasing the authorized speed of the vehicles (existing traffic and weather conditions dictate a safe operating speed).
- Synchronizing delivery and pickup schedules to various units.
- Reducing turnaround time.

b. Types of Haul

(1) Local (Short) Hauls. The ratio of running time to loading and unloading time is brief for local hauls. Trucks running local hauls make several trips per day. The measure of effectiveness for evaluating local haul operations is the amount of tonnage moved during the specified period.

(2) Line (Long) Hauls. The ratio of running time to loading and unloading time is large for line hauls. Trucks running line hauls make only one trip or portion of a trip per operating shift. The measures of effectiveness for evaluating line haul operations are the time consumed, distance traveled, and tonnage hauled during the operational period. The transportation agency expresses this measure in either ton-miles or ton-kilometers.

(3) Zonal (Ring Routes) Hauls. Truck operations confined within the territorial boundaries of one command are intrazonal. Trucks crossing boundaries and operating under the area control

of more than one command are interzonal. The MAGTF commander makes policies and maintains control over interzonal operations, unless the MAGTF is part of a larger effort. Zonal hauls could be a subset of local and line hauls.

c. Hauling Methods

(1) Direct Haul. A direct haul completes a single transport mission in one trip. No transfer of supplies or exchange of equipment occurs. The commander uses direct haul to speed forward movements before establishing transfer or exchange points. This method is most common for local hauls because long-distance direct hauls are hard on both the driver and equipment.

(2) Shuttle (Round Robin). A shuttle involves the same vehicles making repeated trips between two points. This method is most common for local hauls.

(3) Relay. Relay hauling is the continuous movement of supplies or troops over successive segments of a route without transferring the load. The motor transport unit does a relay by changing drivers, transportation equipment, or both for each segment. This method is most common for line hauls. The relay system using transportation equipment is the most-efficient method of line-haul operations. This technique is best used when there is a well-developed road network that is not subject to interdiction. Relay is also the best method to use when the unit cannot complete a one-way haul in one day. Containerization increases the effectiveness of this system by making better use of the truck's tonnage capability. This system provides rapid throughput of cargo and guarantees adequate supervision and support along each segment of the route.

d. Cargo Throughput

Clearing cargo from a beach, port, railhead, and airfield permits continuous discharge of ships, trains, or aircraft. Terminal operation units are responsible for cargo clearance. The availability and proper use of motor transport and MHE are essential. The transportation support unit plans and sets up the circulation network and regulates the flow of vehicles throughout the terminal area. Beach clearance operations are especially difficult as a result of the generally poor road conditions and the temporary nature of the available support facilities. Air terminal clearance is easier because roads and facilities are often better; however, in an effort to unload the aircraft and clear the terminal rapidly, vehicles may not be loaded to maximum capacity.

e. Convoy Operations

Convoys are task-organized to meet the requirements of the assigned mission. A convoy may include a transport element, escort or security element, C2 element, and various support elements. Units plan and execute their own convoy operations. The convoy commander is the direct representative of the commander initiating the operation and is responsible for the conduct, safety, security, and accomplishment of the convoy's mission. Higher headquarters often establish control measures and regulations governing convoy operations on main supply routes. Commanders publish control measures and regulations in local SOPs and in their OPORDs. These control measures include start points, checkpoints, halts, and release points. Area commanders also classify routes in their AO.

f. Types of Routes

(1) Open Route. This is the least-restrictive control measure. Any unit may use the route without a convoy clearance or request. Minimum control is exercised.

(2) Supervised Route. The Marine air-ground task forced movement control center (MMCC)/unit movement control center (UMCC) will specify the size of convoys, the type of traffic, or characteristics of vehicles that require a convoy clearance to use the route. Limited control is exercised.

(3) Dispatch Route. A ground transportation request is required to use this route regardless of the number or types of vehicles. A dispatch route designates when traffic volume is expected to exceed capacity or when the route is critical to operations and priority of use is strictly enforced. Full control is exercised.

(4) Reserve Route. This route is reserved for the exclusive use of a particular unit or type of traffic, and no other units/traffic may use the route. Reserved routes may be identified for large unit movements. Examples include battle handovers, passage of lines, and commitment of the reserve or withdrawals.

(5) Prohibited Route. This route is closed and no unit/traffic may use it. Prohibited routes may result from washouts, destroyed bridges, threat considerations, construction work, or the intent to not interfere with local peak traffic congestion. Availability depends on the nature of the closure (e.g., repairs from battle damage).

2302. PORT AND TERMINAL OPERATIONS

a. Ship-to-Shore Movement
This type of movement is in the assault phase of the amphibious operation that includes the deployment of the landing force from the ships to designated landing areas.

b. Shore-to-Shore Operation
This assault operation moves personnel and materiel directly from a shore staging area to the objective. It does not involve further transfers between types of craft or ships incident to the assault movement. Usually a single-Service operation, a shore-to-shore operation involves water crossings in assault craft and/or aircraft. The purpose of this operation is to establish a force on, or withdraw it from, the far shore.

c. Logistics Over-the-Shore
Logistics over-the-shore (LOTS) operations are the loading and unloading of ships without the benefit of deep draft-capable, fixed port facilities or as a means of moving forces closer to tactical assembly areas dependent on threat force capabilities. The LOTS operation may be conducted over unimproved shorelines, through partially destroyed ports, shallow-draft ports, and ports that are inadequate without LOTS capabilities.

LOTS operations are used to load and unload—

- Break bulk ships.
- Roll-on and roll-off ships.
- Container ships.
- Bulk petroleum, oils, and lubricants ships.
- Water ships.
- Barges.

d. Joint Logistics Over-the-Shore

Joint logistics over-the-shore (JLOTS) operations may involve units and equipment from the Army, Navy, and Marine Corps and may follow amphibious assault operations. The transition from amphibious to JLOTS operations entails passing command of shore facilities to the Army once the amphibious operation ends. The JTF or unified commander directs such transitions. Amphibious operations and MPF operations use some of the same equipment and procedures as JLOTS operations. See JP 4-01.6, *Joint Tactics, Techniques, and Procedures for Joint Logistics Over-The-Shore (JLOTS),* for a detailed discussion of JLOTS operations.

e. Inland Waterway Operations

An inland waterway normally operates as a complete system. It involves rivers, lakes, canals, intra-coastal waterways, and two or more water terminals. Inland waterways can relieve pressure on other modes of transportation. They are especially useful for moving a large volume of bulk supplies and heavy-outsized items that are not easily transported by other means. Although economical, inland waterways are relatively slow compared to other means of transportation. They are especially vulnerable to enemy action and climatic changes.

f. Inland Terminal Operations

Inland terminals serve air, rail, and motor transport operations. They provide cargo transfer facilities at interchange points. They form connecting links when terrain and operational requirements cause a change in carrier.

g. Staging Area Operations

Marine air-ground task forces conduct staging area operations during amphibious and other types of movements:

- Staging area operations for amphibious or airborne movements are between the mounting area and the objective of an amphibious or airborne expedition, through which the expedition, or parts thereof, passes after mounting for refueling; regrouping of ships; and/or exercise, inspection, and redistribution of troops.
- Staging area operations for other movements are established for the concentration of troop units and transient personnel between movements over the lines of communication (LOC).

2303. AERIAL DELIVERY OPERATIONS

Aerial delivery offers the LCE options for supply operations that present potential economies in terms of responsiveness, assets, and security. Aerial delivery lends itself to supply support operations in verticalborne and subsequent operations ashore, especially for bulk items (i.e., Classes I, III, and V). As the initial resupply effort in support of verticalborne operations, coordinated aerial delivery operations can reduce ground transportation requirements while enhancing the sustainability and combat power of the supported force. As a means of sustainment in subsequent operations ashore, aerial delivery can reduce the vulnerability of resupply convoys to enemy interdiction. In this case, economy of effort is achieved through the compensatory reduction of security requirements associated with aerial delivery.

2304. DEPLOYMENT

All operating forces, bases, stations, and depots, except Marine Corps Recruiting Command, may be involved with MAGTF deployments. All MAGTFs deploy from their permanent installations for forward deployments and combat operations. Regardless of the type of deploying force, designated transportation operating agencies control and coordinate the marshaling, embarkation, and movement of the forces.

a. External Transportation Agencies

The following commands are external to the Marine Corps and may be involved with MAGTF deployments:

- Supporting and supported geographic combatant commanders.
- Fleet commanders.
- Defense Logistics Agency (including remote storage activities).
- United States Transportation Command and its subordinate commands:
  - Military Sealift Command.
  - Air Mobility Command.
  - Surface Deployment and Distribution Command.

b. Modes of Transportation

Transportation modes vary depending on the type of MAGTF, the purpose and duration of the deployment, and the anticipated employment. Deployments of larger MAGTFs require use of several transportation modes.

(1) Amphibious. Amphibious deployments can consist of the following modes of transportation:

- Military or commercial trucks, buses, and rail from origins to ports of embarkation and from ports of debarkation to final destination for all personnel, supplies, and equipment.
- Amphibious ships from SPOE to the operating area.
- Air Mobility Command or commercial charter airlift for FIE, follow-on echelon, and replacement personnel who cannot deploy by ship.

- Flight ferry of ACE aircraft.
- Commercial ships from SPOEs for the FIE.

(2) Maritime Prepositioning Force. A MPF deployment can consist of the following modes of transportation:

- Military or commercial trucks, buses, and rail from origins to ports of embarkation(POEs) for personnel, supplies, and equipment.
- Military or commercial air from origins to POEs for personnel, supplies, and equipment (i.e., flying in echelon).
- Flight ferry of self-deploying ACE aircraft.

(3) Marine Expeditionary Force. The MEF deployments are the most complex deployments from a transportation perspective. The MEF elements deploy from different bases and stations that may be in widely separated geographic areas. A forward-deployed MAGTF may be present and serve as an enabling force as additional MEF forces deploy.

(4). Forward-Deployed Marine Air-Ground Task Forces. Forward-deployed MAGTFs will deploy aboard amphibious ships or a combination of air and MPS ships for MPF operations. Transportation support planning frequently requires coordination with HNS or military detachments at foreign ports and airfields to arrange augmentation by foreign civilian transport and US common-user, land-transportation agencies.

2305. Employment

Transportation available for employment in theater includes the organic assets of the MAGTF. It may also include transportation belonging to the joint force commander or to the HN. Specific capabilities depend on the situation. Transportation assets may include airlift, rail, trucks, ships, boats, barges, and pipelines.

The MAGTF commander is responsible for movement control in the MAGTF operating area. Normally, the commander delegates this responsibility to subordinate commanders within whose zones of action or areas the movement takes place. Behind the GCE rear boundary, this normally is the LCE commander.

When operating as part of a joint, allied, or coalition force, the MAGTF commander follows the traffic management and movement control regulations of that command. Normally, the higher commander establishes a movement control agency to provide movement management services and highway traffic regulation. This agency coordinates with allied and HN movement control agencies. See ATP 4-16, *Movement Control,* for a discussion of movement control in a theater of operations.

The MAGTF should use HN transportation support to augment its organic transportation capabilities. Upon arrival in theater, MAGTF civil affairs units should investigate the availability of such support. When operating in NATO or other coalition countries, the MAGTF is obligated to abide by certain agreements among the participating nations. These agreements are called standardization agreements (STANAGs) in the NATO arena and quadripartite standardization agreements (QSTAGs) in the American, British, Canadian, Australian, and New Zealand (commonly referred to as ABCA) Armies Program arena.

SECTION IV. GENERAL ENGINEERING

The term "general engineering" describes those engineering capabilities and activities, other than combat engineering, that modify, maintain, or protect the physical environment. General engineering is applicable to all MAGTFs conducting missions across the full range of military operations. It encompasses a wide spectrum of activities, including horizontal and vertical construction, explosive ordnance disposal, hygiene services (i.e., showers/laundry), tactical bridging, reconnaissance, bulk water production, and bulk fuel storage. For detailed discussions of all aspects of general engineering, see MCWP 3-17.7.

Considering its relative importance to support sustained operations, each element of the MAGTF possesses an organic capability to perform general engineering tasks. Within the GCE, the CEB is capable of providing limited general engineering to the GCE. This can include providing mobile electric power to meet the needs of the Marine division headquarters. Within the ACE, the MWSS is capable of providing limited general engineering support to meet the direct needs of a MAG/ACE. This support can be provided at a forward operating base (FOB) or at a FARP. Multiple MWSSs can be deployed to support the deployment of multiple MAGs.

Within the LCE, the ESB provides general support and general engineering across all elements of the MAGTF. Under certain operational conditions, the MAGTF may request and receive augmentation from the NCF to address a shortfall of organic general engineering capacity or to provide a unique capability (e.g., well drilling) that does not organically reside within the MAGTF. The NCF augmentation can include a single NMCB or multiple NMCBs. See MCWP 4-11.5 for a more comprehensive description of the command relationships regarding these augmentation units.

Due to scope and scale, general engineering activities are heavily reliant upon transportation, supply, maintenance, and funding. Detailed planning, coordination, and prioritization are performed at the MAGTF headquarters by the MAGTF engineer staff officer in conjunction with engineer staff officers at the HQ of each combat element (i.e., GCE, ACE, and LCE). This planning and coordination ensures that sufficient ground transportation is allocated to move heavy, oversized construction equipment, personnel, and supplies to project sites. Due to the size and cost per item, the MAGTF possesses limited numbers of specialized general engineering equipment. As a result, the readiness of this equipment is critical to maintain the MAGTF's organic general engineering capacity. Staff coordination is also provided to ensure that the large quantities of bulk materials (e.g., gravel, sand, lumber) arc procured and delivered when they are needed at each construction site.

2401. Engineering Tasks

Engineering tasks range from support provided by Marine engineer organizations to external support provided by assigned forces such as the NCFs and civilian or host nation resources. The subfunctions of general engineering encompass several tasks, many of which might also be described as combat support tasks. Table 2-2, on page 2-29, shows a wide range of engineering tasks assigned to engineer organizations.

2402. Marine Air-Ground task force Engineering Unit Missions

a. Combat Engineer Battalion

The CEB mission is to enhance the mobility, countermobility, and survivability of the MARDIV. The battalion consists of an H&S company, three combat engineer companies, an engineer support company, and a mobility assault company. The battalion is capable of providing task-organized platoon-based detachments to battalion landing teams and reinforced combat engineer companies to regimental landing teams. The remainder of the battalion is capable of providing general support combat engineering to meet the requirements of the GCE of a MEB-sized MAGTF. Within a MEF-sized MAGTF, the battalion would fall under the command and control of the MARDIV HQ. See MCWP 3-17.7 for further description of the capabilities and organization of the CEB.

b. Engineer Support Battalion

The ESB mission is to provide general engineering support of an expeditionary nature to the MEF. This includes survivability, countermobility and mobility enhancements, EOD, and general supply support handling, storage, and distribution of bulk Class I (water) and bulk Class III (fuel). The battalion consists of an H&S company, two engineer companies, engineer support company, bulk fuel company, EOD company, and bridge company. The battalion is capable of providing task-organized detachments to MEUs and CLBs, and direct support to the HQ regiment. The remainder of the battalion is capable of providing general engineering support to meet the requirements of a MEB-sized MAGTF. In this operational environment, the battalion would fall under the command and control of the LCE HQ regiment. Within a MEF-sized MAGTF, the battalion would fall under the command and control of the MLG HQ. For further discussion, see MCWP 3-17.7 for the capabilities and organization of the ESB.

c. Marine Wing Support Squadron

The MWSS mission is to provide aviation ground support to enable a composite MAG and supporting or attached elements of the MACG to conduct expeditionary operations. The squadron consists of an H&S company, airfield operations company, engineer company, and motor transport company. The engineer company consists of combat engineer platoon, utilities platoon, and heavy-equipment platoon. A bulk-fuel platoon and the EOD section are located within the airfield operations company. For further discussion, see MCWP 3-21.1 for the capabilities and organization of the MWSS.

Table 2-2 Engineering Task Matrix.

Tasks	Combat Engineer Battalion	Engineer Support Battalion	Marine Wing Support Squadron	Naval Construction Force	Civilian/Host Nation Support
Beach improvements		X		X	
Camp construction, repair, and/or maintenance		X	X	X	X
Construction design		X		X	X
Demolition	X	X	X	X	
Engineer reconnaissance	X	X	X	X	X
Explosive ordnance disposal		X	X		
Field fortifications	X	X	X	X	X
Obstacle removal	X	X	X	X	X
Pioneer roads	X	X	X	X	
Planning and installation of obstacles and/or barriers	X	X	X	X	X
Pre-engineered structures		X	X	X	X
Rapid runway repair		X	X	X	
Tactical water and/or hygiene service		X	X	X	
Tactical bulk fuel storage		X	X		
Tactical electrical supply		X	X	X	
Unpaved roads, airstrips, and/or marshaling areas		X	X	X	X
Vertical takeoff and landing and/or helicopter landing zone		X	X	X	X
War damage repair		X	X	X	X

2403. NAVAL CONSTRUCTION FORCE

a. Naval Construction Force

The Naval Construction Force (NCF), or Seabees, provides a wide range of general engineering support to the MAGTF and joint force and is comprised of six primary agencies: the naval construction group (NCG), the naval Ccnstruction regiment (NCR), the naval mobile construction battalion (NMCB), the amphibious construction battalion (ACB), the underwater construction team (UCT), and the construction battalion maintenance unit (CBMU). Collectively, the NCF is staffed, structured, and equipped to perform limited combat engineering, heavy general engineering, and heavy construction support to the MAGTF or joint force in permissive, uncertain, and hostile environments. Seabees perform general engineering tasks at the initial, temporary, semi-permanent, and permanent construction standard primarily focusing on large-scale construction, maintenance, and repair projects to include roads, bridges, airfields, base camps, and logistic sites as well as provision of power generation and distribution, water well drilling, quarry operations, and base camp operation and maintenance. Seabees reinforce and augment Marine Corps engineer capabilities, enhance the MAGTF's ability to conduct civil-military operations and support specialized engineering and construction capabilities not resident within the MAGTF. For further information see MCWP 4-11.5.

b. Naval Construction Regiment

The NCR is an independent command element that conducts construction and engineer project management operations. It provides C2 over assigned subordinate engineer units and other expeditionary units. When employed with subordinate units, an NCR is a flexible organization structured to accomplish the full range of general engineering tasks and limited combat engineering tasks in a mid- to high-level intensity conflict. It may provide general engineering C2 at the MEF level and provides a range of engineering expertise from contingency planning through force projection to a fully developed MEF AO.

The NCR contains specialized engineer units, construction equipment, professional expertise, and C2 assets required to support MEF operations. The mix and type of units attached to the NCR are determined by the supported command's organization and mission. The NCR consists of a CE, multiple task-organized Seabee units, and may have other Service or HN engineer units under its control. Subordinate units include—

- NMCB.
- CBMU.
- UCT.
- ACB.

c. Naval Mobile Construction Battalion

The NMCB is the primary Seabee unit for conducting construction and engineer operations. Its personnel and equipment are a modular task organization of air-transportable, ground, and sea logistic elements. The NMCB is the NCF's basic operating organization. The NMCBs can deploy rapidly as part of expeditionary ready forces, MPFs, and air contingency forces. Direct labor assets are approximately seventy percent of the total basic allowance that can be assigned directly to construction and contingency operations or other tasking.

d. Construction Battalion Maintenance Unit

The CBMU provides deployment support for First Naval Construction Division, conducts public works functions at expeditionary FOBs, and constructs expeditionary medical facilities. The mission of a CBMU is to rapidly deploy flexible and versatile detachments to provide responsive general engineering support to forces ashore during contingency operations.

e. Amphibious Construction Battalion

The ACB conducts STS transportation of bulk fuel/water supplies, materials, and equipment in support of amphibious operations known as JLOTS operations. They provide support to MPF operations and conduct related CSS, including general engineering and security support in subsequent operations ashore. The ACB is organized under the NBG as operational commands.

2404. JOINT ENGINEERING / INTERAGENCY ENGINEERING

a. United States Army Engineering

Embedded engineer capabilities in brigade combat teams enable rapid response and fight upon arrival. Embedded engineers provide the capability for mobility that units require full time. They are uniquely organized and equipped based on the type of parent brigade combat team (armored, infantry, or Stryker). They maintain an engineer force pool of readily accessible capabilities, with a C2 HQ that supports division and theater HQ with the versatile engineer support needed for a campaign-quality Army. Headquarters will either be designated as combat effects or construction effects.

b. United States Air Force Civil Engineering

United States Air Force civil engineering provides a full spectrum of engineering support to establish, operate, and maintain garrison and contingency airbases. They provide fire protection services; expedient construction; chemical, biological, radiological, and nuclear protection; and explosive ordinance demolition. The Air Force's civil engineers also establish beddown facilities; sustain airfield operations; and perform water-well drilling, airfield pavement evaluations, quarry operations, materiel testing, expedient facility erection, and concrete and asphalt paving.

SECTION V. HEALTH SERVICE SUPPORT

Health service support is a process that delivers a healthy, fit, and medically ready force; counters the health threat to the deployed force; and provides critical care and management for combat casualties. The focus of HSS emphasizes the provision of far-forward deployed, mobile medical, and surgical support; with an assigned mission to triage, treat, evacuate, and return to duty US tactical forces. For further guidance, see MCWP 4-11.1 and JP 4-02, *Health Service Support*.

2501. ROLE VERSUS ECHELON (LEVEL) OF CARE

"Echelon" or "role" is defined on the basis of capabilities and resources, and is not specific to particular medical unit types. The term "role" is a NATO terminology used by land or air forces, while "echelon" is primarily a maritime term.

While closely related, the terms "role" and "level" are not exactly interchangeable. "Roles" have four echelons of medical care, while "levels" have five echelons of medical care: Levels I, II, III, IV (which is outside the continental United States medical treatment facility), and V (which is CONUS definitive care). At the tactical level, medical support will be provided at Level I and Level II. Tactical-level medical personnel consistently coordinate with higher echelons of care concerning other functions of HSS (e.g., reports submission, patient movement, and Class VIII resupply). For further guidance, see *NATO Logistics Handbook*.

The tactical levels of medical care are—

Level I
- First responder.
- Pre-hospital treatment.
- Damage control resuscitation.
- Examples: self-aid/buddy aid, corpsman, shock trauma platoon.

Level II
- Damage control surgery.
- Limited ancillary services.
- Limited diagnostic capability.
- Limited patient holding.
- Examples: forward resuscitative surgery system, fleet surgical team, CRTS.

2502. MARINE AIR-GROUND TASK FORCE CAPABILITIES

Tactical-level MAGTF forces are deployed as a MEF, MEB, MEU, or SPMAGTF. Dependent on the size, mission, and physical characteristics of the AO, each of the attached commands and combat elements of the deployed force will deploy with an appropriate medical capability. Any assigned medical personnel are commonly referred to as a HSS. Tactical logistics includes organic unit capabilities and the combat service support activities conducted in support of military operations. The goal of tactical-level HSS is to support maneuver forces in the battlespace.

a. Command Element
The CE of the MAGTF is capable of coordinating routine and emergency treatment and preparation for evacuation by using its organic medical section. These HSS functions are normally performed by a subordinate battalion/regimental aid station.

b. Ground Combat Element
The GCE is the most forward fighting force and has a higher exposure to combat wounds and casualties. Injured and sick personnel requiring hospitalization are readied or evacuated along the continuum of care as their needs dictate. Normally, a regimental or battalion aid station serves as the hub for medical support. See MCWP 4-11.1 for a detailed discussion.

Headquarters battalion, Marine division, medical section, and regimental and battalion infantry unit medical platoon/section provides—

- Preventive medicine.
- Treatment for minor illnesses and injuries.
- Emergency lifesaving for battle and nonbattle casualties.
- Emergency treatment and preparation for evacuation of all casualties.
- Disease prevention and control measures supervision.

c. Aviation Combat Element
Health services personnel are assigned to the primary subordinate organizations in the MAW. At the EAF the MWSS provides aid station capability and personnel. Squadron unit medical platoon or section provides—

- Preventive medicine.
- Treatment for minor illnesses and injuries.
- Emergency lifesaving for battle and nonbattle casualties.
- Emergency treatment and preparation for evacuation of all casualties.
- Specific aerospace medicine needs for treatment and procedures.

d. Logistics Combat Element
The MLG provides direct and general medical support to supported units in excess of their assigned or organic capabilities. The MLG serves as the hub for the combatant surgeon and patient regulation services.

e. Medical Battalion
The medical battalion's primary mission is to perform those emergency medical and surgical procedures that, if not performed, could lead to loss of life, limb, or eyesight. The medical battalion is made up of an H&S company and two or three surgical companies, according to parent MEF force structure. The H&S company and the surgical companies contain a varying number of surgical platoons, which are comprised of shock-trauma platoons, forward resuscitative surgical systems, en route care systems, patient holding wards, and ancillary services. For additional guidance on patient evacuation see MCRP 4-11.1G, *Patient Movement*.

The battalion's surgical companies provide the following support:

- Initial trauma resuscitation and surgical intervention.
- Temporary casualty holding.
- Ancillary services (laboratory, radiology, pharmacy, and combat stress).
- Medical regulating.
- Ground evacuation support to forward medical elements.
- Limited and task organized en route care evacuation support from forward surgical or treatment elements to shore or sea-based medical treatment facilities for critically ill or injured patients by opportune ground, air, or sealift.
- Preventive medical support.

f. Medical Logistics Company
Medical supplies and equipment (Class VIII) for the MEF are managed through the medical logistics company (MEDLOGCO), which issues the AMAL and ADAL and handles resupply issues. The MEDLOGCO is directly responsible to the supply battalion commanding officer supporting the medical battalion. The MEDLOGCO conducts the following functions:

- Maintains medical equipment.
- Maintains centralized acquisitions, storing, and stock rotation.
- Constructs medical supply blocks.

g. Dental Battalion

The dental battalion task-organizes dental sections and detachments to HSS elements of the MAGTF. In an operational environment, the dental battalion's primary mission is to provide dental health maintenance with a focus on emergency care. In addition to medical support determined appropriate by medical battalion and surgical company commanders, dental detachment personnel may provide the following support:

- Casualty collection and clearing casualty triage.
- Mass casualty evacuation.
- Postoperative care.
- Holding ward care.
- Central sterilization.
- Supply room.

2503. CAPABILITIES EXTERNAL TO THE MARINE AIR-GROUND TASK FORCE

In a purely expeditionary theater, the Marine Corps/Navy team does not have all the organic capabilities to meet the full spectrum of HSS requirements; therefore, there is always, as with all logistics, the need for joint/inter-Service relationships. The medical capabilities external to tactical-level logistics include Levels III, IV, and V echelons of medical care. These assets and capabilities include—

- **Expeditionary Medical Facilities.** Medical systems that are deployable to an area involved in tactical operations. Expeditionary medical facilities provide Level III capabilities to include an increased bed count (up to 150 beds) and evacuation capabilities.
- **Hospital Ships.** The hospital ship is a Level III, 1,000-bed floating surgical hospital. Its mission is to provide acute medical care in support of combat operations at sea and ashore. See JP 4-02, *Health Service Support*, for more information.
- **Casualty Receiving and Treatment Ships.** The CRTSs are a Level II facility of amphibious ships in the ATF. For medical support capabilities of these vessels and their potential roles as CRTSs review JP 4-02.

2504. PATIENT MOVEMENT

Prompt movement of casualties through the evacuation system to treatment facilities is essential to decrease morbidity and mortality of battlefield casualties. A sound patient movement process ensures that patients move along the continuum of care as their needs dictate. This process also ensures the efficient and effective use of limited HSS assets. See MCRP 4-11.1G, for a general summary of the HSS systems and specific tactics, techniques, and procedures for patient movement. For patient movement in joint operations, refer to JP 4-02.2, *Joint Tactics, Techniques, and Procedures for Patient Movement in Joint Operations.* Patient movement involves two distinct components:

- **Evacuation.** During evacuation, patients are moved between point of injury or onset of disease to a facility that can provide the necessary treatment capability.
- **Medical Regulating.** Medical regulating involves the actions and coordination necessary to arrange for the movement and tracking of patients through the levels of care. This process matches patients with a medical treatment facility that has the necessary HSS capabilities. It also ensures that bedspace is available. In the medical regulating phase, destination medical treatment facilities are selected that are equipped with the necessary HSS capabilities for patients being medically evacuated throughout different theaters of geographic combatant commands.

SECTION VI. SERVICES

The various nonmateriel and administrative support activities of the services functions are described in JP 4-0, Naval Doctrine Publication (NDP) 4, *Naval Logistics,* and MCDP 4. The Marine Corps categorizes services functions as being either CSS services or command services.

2601. COMBAT SERVICE SUPPORT SERVICES

a. Disbursing

Manpower restrictions mandate that the committed MAGTF's disbursing support be located in the LCE rear area. Geographical separation of the ACE, GCE, and LCE units necessitates collocating disbursing offices that are capable of providing the required disbursing services to both the ACE and the GCE. These offices respond to the taskings of their respective commanders but receive procedural direction from the MAGTF disbursing officer, who is solely responsible for all disbursing operations.

(1) Deployment Capability. Disbursing assets of the LCE can be deployed to provide full-service disbursing support for all MAGTF organizations. Services for a MEF in theater are provided by the MLG disbursing sections and platoons. This flexibility allows for the task-organizing of disbursing assets to meet the needs of the MAGTF commander.

(2) Phases of Support. Disbursing support meets two primary missions in theater; the payment of MAGTF obligations and pay-related support for deployed Marines and sailors. Disbursing support is divided into three phases:

- **Phase One.** During the initial assault phase, when the force is establishing itself ashore, required disbursing services are minimal. Normally, the capability for payment of MAGTF obligations and/or individual emergency payments to Marines is available. During this phase, mission accomplishment and survival divert attention to the

battlespace and disbursing personnel may be committed to augmenting other LCE efforts. Therefore, a minimum of disbursing and accounting requirements are met.
- **Phase Two.** This phase begins when the need to establish an office to provide increased service is identified. In addition to phase one support, on-call pay support is coordinated. The contact team approach is used to deliver support to MAGTF elements.
- **Phase Three.** The third phase is usually conducted during sustained operations ashore. In addition to disbursing tasks accomplished in phase two, phase three services may include, but are not limited to—
  - Monthly on-call pay support.
  - Guidance to the MAGTF commander on disbursing matters.
  - Public voucher payments for contracts and condolence payments.
  - Data systems input for updating the central file, generating required reports, and submitting financial returns.
  - Cash depository for the Marine Corps Exchange (MCX), postal service, and clubs.
  - Personal and US Treasury check cashing.
  - Foreign currency support.
  - Cross-Service support, as required.
  - Pay agent training and appointments.

b. Postal

Postal assets are task-organized to provide postal support for the MAGTF and attachments. These assets include a mobile main post office and 12 mobile unit post offices. The main post office coordinates all postal functions and locations. Each unit post office is capable of providing full postal support to a reinforced regiment. Smaller detachments can be task-organized to support various-sized MAGTFs.

(1) Support. The bulk of postal support is located throughout the MAGTF rear area. Unit post offices provide postal support to various CSSAs. On request from the GCE, mobile unit post offices may be located in the GCE rear area. The ACE may also request mobile unit post offices. These mobile units can provide full or partial postal services. In the event that postal services are not requested by the GCE or ACE, the mail delivery for GCE rear and ACE personnel is accomplished through resupply channels. All postal units respond to the taskings of their respective LCE commanders but receive procedural direction from the MAGTF postal officer, who is solely responsible for all postal operations.

(2) Phases of Support. During amphibious operations, postal support is divided into three phases:

- **Phase One.** During the assault phase, postal services generally are not available.
- **Phase Two.** This phase begins when the need to establish a postal unit is identified. In addition to processing incoming and outgoing personal and official mail, unit post offices provide all postal services that are normally available in-garrison. Mail deliveries to units are accomplished by unit mail clerks and orderlies.
- **Phase Three.** The third phase begins when sufficient forces are ashore to establish a rear area. In this phase, postal assets are committed in support of the MAGTF mission and perform the following functions:
  - Advise the MAGTF commander on postal matters.

- Route mail to and from the battle area.
- Sell stamps and money orders.
- Accept letters and packages for mailing.
- Deliver and dispatch official and personal mail.
- Establish a casualty mail section.
- Coordinate the resupply of unit postal offices operating throughout the area (unit post offices are stocked with the supplies and equipment to support regimental-sized organizations for a period of 60 days without resupply).
- Coordinate cross-service support, as required.

c. Marine Corps Community Services

Marine Corps Community Services (MCCS) must be included in the initial planning process to ensure that the proper logistic and manning support are identified. Staffing requirements are based on one officer or staff noncommissioned officer and three enlisted Marines to serve 1,000 customers. Two enlisted Marines are added per each additional 1,000 customers or portion thereof. This manning is based on operating a tactical field exchange for nine hours per day, plus an additional three hours per day for restocking and administrative requirements. Additional MCCS Marines will be needed for clubs, food operations, recreation, fitness centers, and morale satellite.

A tactical field exchange is established when no other source of Class VI support is available. The MAGTF commander determines when to establish a tactical field exchange, but the LCE commander designates the site for the exchange. When needed, mobile tactical field exchanges are sent to MAGTF maneuver elements. See MCWP 4-11.8.

The establishment of a field exchange is based on three broad categories of deployment. The three categories and their retail requirements are as follows:

- Category I deployments are exercises with little opportunity for Marines to access MCCS assets. Limited MCX and services operations exist (e.g., 1-day access to retail or exchange facility).
- Category II deployments or exercises up to 90 days in duration, with some opportunities for Marines to access MCCS assets. The MCX and services operations are expanded to include repeated access to the base MCCS or HN facilities. Units may request the establishment of a tactical field exchange.
- Category III deployments or exercises in excess of 90 days in duration, and Marines have the opportunity to access MCCS assets. The MCX and services operations are greatly expanded to include access to base MCCS or HN service facilities.

The LCE provides the supported tactical field exchange units with critical logistic and/or administrative support and services, including the following:

- Facilities to house the tactical field exchange (tents or other structures).
- Class IV supplies.
- Tactical field exchange disbursing.
- Security.
- Transportation and MHE.

- Utilities and communications.
- Services to tactical field exchange personnel (i.e., billeting, subsistence, disbursing, postal, legal, etc.).

The LCE coordinates logistic support for mobile tactical field exchanges, which deliver health and comfort items to troops that otherwise would be unable to access a tactical field exchange due to their mission and geographic location.

(1) Tactical Field Exchange Operations. A deployed tactical field exchange activity is operated as a branch of the parent Marine Corps exchange from which the unit is deployed. All internal supplies, resale goods, and any resupply items are provided from that parent exchange or through a memorandum of understanding with sister Services. In the event of an extended deployment and/or employment or an extensive mobilization, exchange services will be provided by using MCCS nonappropriated funds.

(2) Concept of Organization. Support is provided in the form of a direct operational tactical exchange from a mobile tactical field exchange capability by the LCE. Only Class VI supplies required to stock the field exchange are provided by the MCX.

(3) Resupply. The LCE exchange officer initiates resupply of Class VI supplies for short-term support. Class VI supply items are coordinated and shipped in the same manner as other supply blocks for deploying units.

(4) Logistics Combat Element Functions. The LCE provides exchange support for the MAGTF by providing LCE Marines holding military occupational specialty (MOS) 4130 or 4133 to make up the exchange platoon. The 4130/4133 MOS is composed of sergeants through warrant officers.

(5) Marine Corps Community Services Officer. The MCCS officer advises the commander on all matters pertaining to the MCCS operations. The MCCS officer is accountable for MCCS resources and personnel. The MCCS officer must be involved in the planning phase as early as possible to provide an appropriate level of support. The MCCS officer will serve as the liaison between the command and the authorized support organizations, such as Army and Air Force Exchange Services, Armed Forces Entertainment, Army MWR, and Space and Naval Warfare Systems Command.

(6) Marine Corps Community Services Staff Noncommissioned Officer. The senior enlisted MCCS Marine is designated the staff noncommissioned officer in charge and is essential to the planning, execution, and operation of MCCS services at all levels of the MAGTF. The MCCS Marines are assigned to operate MCX and services, including the direct operational tactical exchange, clubs, food operations, recreation, fitness centers, and morale satellite.

(7) MCCS Civilian Personnel. When the AO has sufficiently matured, and with command approval, the MCCS officer will contact MCCS HQ in Quantico, Virginia, to coordinate manning and deploying MCCS civilian personnel.

d. Security Support

Successful enemy action against C2 facilities and LCE installations can make it impossible for the MAGTF commander to accomplish the assigned mission. Enemy threat, however indirect, may be posed by conventional and/or unconventional forces. Consequently, combat support and LCE installations to the rear of the GCE should be considered high-priority, lucrative targets. The LCE commanders are responsible for the security and survivability of their own units.

(1) Security Measures. All commanders must take both passive and active measures to provide security and to ensure the continuation of their units' missions despite the threat or the initiation of enemy action.

(2) Role of the Military Police. Security support is provided by the military police units at the MEF level and can be task-organized. These units, however, are insufficient to provide all security support functions simultaneously. A military police unit is an economy-of-force unit that must be used wisely. Support is based on the concept of operations and a clear understanding of priorities as established by HHQ. In support of the MAGTF, the military police functions include—

- Antiterrorism and force protection.
- Maneuver and mobility support operations.
- Area security operations.
- Law and order operations.
- Internment operations.
- Processing EPWs.

e. Legal Services

The MLG, H&S battalion, legal services support section is the command entity that provides legal services support for the MAGTF. In support of a MAGTF, legal services support tasks are normally performed by the LCE through one or more legal services support teams.

(1) Legal Services Support Teams. Teams are employed at appropriate times and places in support of major MAGTF personnel concentrations in the AO. Legal services support teams vary in number, size, and composition depending on the—

- Mission, size, and composition of the MAGTF.
- Expected duration of the operation.
- Scheme of maneuver and topography involved in the operation.

(2) MAGTF Support. Most legal services work in support of MAGTF operations involves—

- Injury, death, claims, and supply investigations.
- Legal review of OPLANs.
- Law of war training.
- Rules of engagement.
- Legal assistance (not including contracting support).

- Nonjudicial punishment.
- Summary and special courts-martial.

(3) Marine Expeditionary Force Support. The MEF operations may involve the deployment of all available legal services personnel. Each major subordinate command has an organic staff judge advocate section to ensure the coordination of legal services support for the command and its subordinate organizations. The staff judge advocate provides legal advice to the commander.

f. Civil Affairs Support

Civil affairs are a command responsibility involving those activities between MAGTF elements, civil authorities, and local civilians in the AO. Whether assigned or task-organized, civil affairs units are normally assigned to the MAGTF CE and function under the staff cognizance of the MAGTF G-3/S-3. They assist in planning and conducting MAGTF civil-military operations to implement MAGTF civil affairs missions and goals within the MAGTF AO. For MAGTF operations, civil affairs responsibilities are normally confined to periods of limited duration between the arrival of the first tactical units and the termination of operations or the transfer of responsibility to Army civil affairs units. MAGTF civil affairs activities are normally limited to those minimum essential civil-military functions that are necessary for the accomplishment of the primary mission. *Refer to JP 3-57, Civil-Military Operations,* for detailed guidance on civil affairs.

Civil affairs support is provided by all individuals and elements of the MAGTF to achieve the established civil affairs goals of the command. As a subfunction of services, civil affairs support is included in the six functional areas of CSS. Civil affairs support tasks are largely logistical in nature and generally involve population and resource control assistance in support of MAGTF operations; however, the capability to perform those tasks is not unique to the LCE. The civil affairs function is conducted in all phases and in every geographic zone of the operation. Supported units also possess civil affairs support capabilities, and the LCE provides support beyond the supported units' organic capabilities. Marine Corps civil affairs assets are MAGTF assets. Current Marine Corps civil affairs units reside in the Reserve establishment. Refer to MCRP 3-33.1A, *Civil Affairs Tactics, Techniques, and Procedures*, for additional guidance on civil affairs.

g. Mortuary Affairs

The DODD 1300.22, *Mortuary Affairs Policy*, states that the human remains of all members of the Armed Forces of the United States will be returned for permanent disposition in accordance with the decedent's will or the laws of the state (territory, possession, or country) of the decedent's legal residence as directed by the person authorized to direct disposition of human remains.

The expedient and respectful repatriation of deceased personnel to the person authorized to direct disposition of human remains is the top priority of the mortuary affairs program. Inherently, every small unit leader and commander bears some responsibility for providing mortuary affairs services. As such, the process begins at the point at which a Marine or Sailor dies. Thus, formal chains of evacuation and accountability begin at the unit level with the establishment of unit search and recovery teams to ensure that this service is properly addressed, and deceased service members are handled with care, dignity, and respect. Unit-level mortuary affairs operations consist of the initial search, recovery, and evacuation of unit deceased personnel to the nearest mortuary affairs collection point.

The Marine Corps has one unit, Personnel Retrieval and Processing Company, resident within the reserve component that provides mortuary affairs services for the MAGTF in support of combat or contingency operations in order to expedite the recovery, processing, and evacuation of human remains to designated facilities. The company can task-organize to form scalable units as required and is capable of deploying in support of combat maneuver units. Refer to JP 4-06, *Mortuary Affairs*, for detailed guidance on Mortuary Affairs. Personnel Retrieval and Processing Company capabilities include the following functions:

- Search and recovery operations.
- Mortuary affairs collection point operations.
- Interment/disinterment operations.
- Mass fatality/casualty incident support operations.
- Theater-level mortuary affairs support operations.
- Mortuary affairs support to disaster relief operations.

h. Contracting Support Services

Because of the unique nature of contracted support in military operations, the Marine Corps has improved our capability to plan and effectively integrate OCS into all phases of MAGTF operations. While OCS includes the legal mechanism (contracting and funding authorities) to obtain commercially provided supplies, services, and construction in support of military operations; the overall responsibility for mission accomplishment remains with the commander, not with the supporting contracting activity.

Due to the small number and high demand of Marines in the contingency contracting force for contracted support in the operational environment, the Marine Corps must ensure that the planning, execution, and management of OCS is optimized and focused on MAGTF capabilities and support. As the Marine Corps places greater emphasis on expeditionary concepts, Marines at all echelons will rely on OCS as an integral component to their operations.

(1) Command and Contracting Authority. Contracting authority is unique within the Marine Corps, in that, it is not granted by a unit commander but originates from the head of the contracting agency. To maintain the contracting authority and command relationship, contracting officers must be placed within the organization where they can maintain functional independence. See figure 2-11 on page 2-42.

(2) MEF Expeditionary Contracting Platoon. The Marine Corps maintains an organic contingency contracting force capability within each MEF. Each MEF has an expeditionary contracting platoon (ECP) within the MLG. The MLG employs its contracting capability as an organic contracting support element to the MAGTF or smaller deploying units to support the numerous operational missions assigned to the Marine Corps. The mission of the ECP is to develop, train, and sustain the appropriate mix of contingency contracting force Marines, to provide responsive expeditionary contracting support, and maintain the technical proficiencies required for the employment as a MAGTF capability. The ECP consist of uniformed personnel

that can rapidly deploy and provide expeditionary contracting support to any size MAGTF or deploying unit during exercises, contingency, and humanitarian assistance disaster relief operations. Marine Corps contracting capability is small but it is scalable to the mission and size of the MAGTF or unit it supports. The ECP is the cornerstone of contingency and expeditionary contracting support for the MAGTF.

(3) Organization. The Marine Corps contracting capability is organized to best support the MAGTF's mission while maintaining control and oversight of the contracting function. See figure 2-12 on page 2-43.

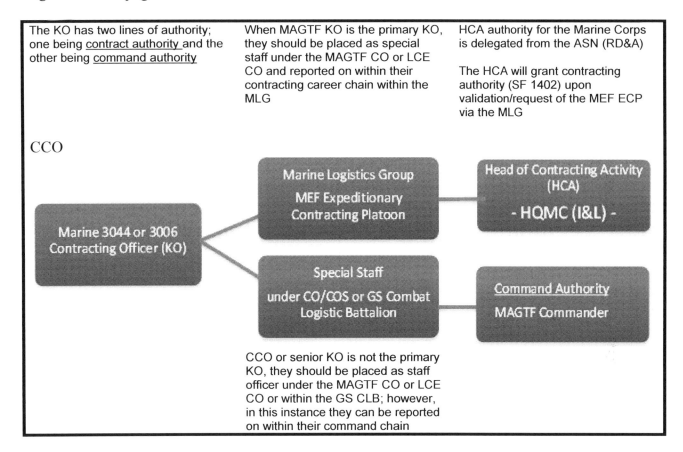

Figure 2-11. Command vs. Contracting Authority (Functional Independence).

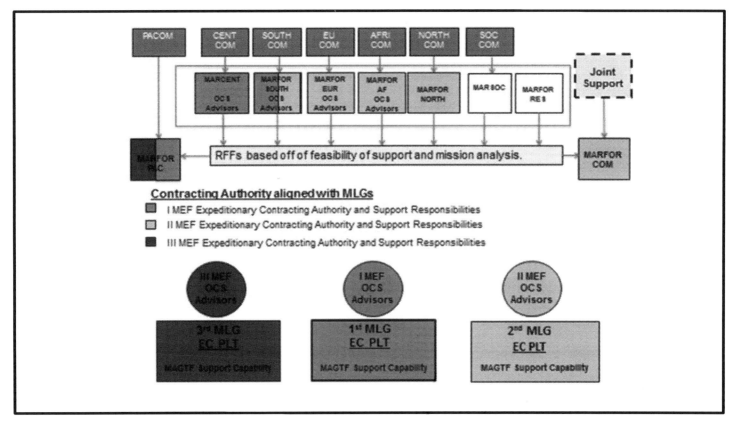

Figure 2-12. OCS Approach to Planning and Integration.

(4) Contracting Support Execution. There are four primary phases for the execution of contracting support during MAGTF operations:

- **Phase 1.** Mobilization/initial deployment.
- **Phase 2.** Arrival in theater/build-up.
- **Phase 3.** Sustainment.
- **Phase 4.** Redeployment and retrograde.

Proper execution of contracting support during each phase by the contracting officer is critical to supporting the MAGTF Commander and helps to ensure mission success. While there are several steps in the acquisition process and processes can vary dependent upon the environment, the diagram in figure 2-13, on page 2-44, illustrates the basic execution cycle for the identification and procurement of a requirement.

As discussed earlier a contracting officer receives a contracting warrant/authority from the HCA. A contracting officer can only request a change to their warrant or contract authority through their parent ECP and MLG. Additionally, the ECP and MLG will be responsible for any required reach-back support, administrative and legal support, contracting operating systems support, or other support as may be required for the contracting officer to successfully execute contracts.

MCTP 3-40B. Tactical-Level Logistics

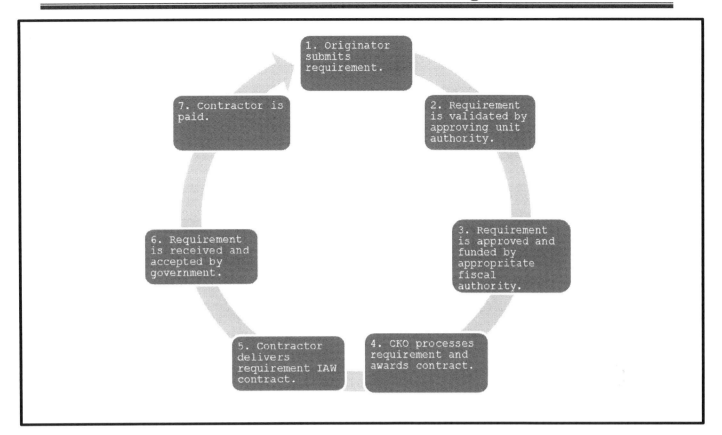

Figure 2-13. Basic Process Diagram for Executing Contract Support.

(5) Micro Purchase Authority. Micro-purchase authorities that are available for consideration to the MAGTF commander are those of the Government-Wide Commercial Purchase Card (GCPC) and the field ordering officer (FOO).

A GCPC is similar in nature to a personal commercial credit card, except it is issued to authorized civilian and military government employees only, for use to acquire and to pay for commercial off the shelf supplies and services to meet the minimal needs of the government. These cards are limited to the same thresholds established as the micro-purchase rules unless the cardholder is otherwise approved and appointed in writing on a Standard Form 1402.

A FOO is an individual who has been authorized by the contracting officer, in writing, to execute micro-purchases in an expeditionary environment by using Standard Form (SF) 44, *Purchase Order-Invoice-Voucher*. The MAGTF commander cannot appoint or delegate the duties of a FOO. The FOOs are employed in expeditionary and contingency environments when MAGTF operations will be spread over a large area of responsibility and the contracting officer determines that the requirements of a specific location or unit do not warrant the deployment of a contracting officer and can be met with the training and appointment of a FOO. The FOOs are commonly used in cases where the unit is operating in a remote area away from the contracting office and GCPC is not the preferred method to execute micro-purchases, because only cash/local currency is the accepted method of payment.

2602. COMMAND SERVICES

a. Personnel Administration

Personnel administration is an important command service conducted at all major levels of the MAGTF. While providing personnel administration is a responsibility of the commander, this function is typically executed under the cognizance of the unit adjutant (G-1/S-1). The G-1/S-1 takes the lead in coordinating action between other staff functions (e.g., G-2/S-2, G/-3S-3, G-4/S-4). It is also responsible for all unit personnel matters, to include the following personnel-related functions:

- Mortuary affairs.
- EPW handling procedures.
- Civilian personnel matters (e.g., contractors, civilian employees, refugees).
- Manpower and personnel management.
- Discipline, law, and order.
- Casualty tracking.

The G-1/S-1 is responsible for preparing Annex E to the MAGTF OPORD, which sets forth the personnel requirements for the MAGTF. This document provides higher and subordinate HQ with a general understanding of how personnel support will be provided for the MAGTF. Normally, Annex E is prepared only at the MAGTF and HHQ level. Annex E should address coordination and support with agencies external to the MAGTF. It should also address any inter-service support or host country agreements. The following areas should be addressed in Annex E:

- Relationships with the International Red Cross.
- Arrangements for transfer of prisoners of war between Services or acceptance of prisoners of war from Allied forces.
- Reports of law of war violations.
- Currency and credit controls.
- Use of US citizen civilian personnel.
- Administration of non-US citizen labor.
- Joint replacement depots.
- Provision of common-user MWR services and facilities.
- Provision of postal and courier services.

b. Religious Ministries Support

Religious ministries support performs ecclesiastic functions and provides both faith-based and nondenominational counseling and guidance for all personnel. It is a significant factor in building and maintaining morale. Chaplains (ordained or accredited priests, ministers, and rabbis) are assigned throughout the MAGTF at the organizational level and higher. Chaplains normally report directly to the commander. For more information, reference MCRP 6-12C, *The Commander's handbook for Religious Ministry Support,* for religious ministry support information.

c. Financial Management

The Marine Corps founded its philosophy of financial management on the principle that financial management is inseparable from command. The commander must make vital fiscal decisions and keep financial management in proper perspective as a part of balanced staff action. In this regard, the commander should recognize that financial management has no bearing on the determination of mission, but rather is a primary consideration in determining both the means and the time-phasing of mission accomplishment. The commander has two types of financial responsibility:

- **Command.** Command financial responsibility parallels the commander's other responsibilities. The commander is tasked with controlling and administering funds granted to perform the mission.
- **Legal.** When in receipt of an allotment or operating budget, the commander is legally responsible for the proper receipt and obligation of those appropriated funds.

Financial management operations within the operating forces are divided into four fundamental areas: budgeting, accounting, disbursing, and auditing. To assist the commander in the accomplishment of these functions, a general staff-level financial organization, the comptroller, is established at each major command. Commanders at lower echelons normally assign the additional duty of fiscal officer to a special staff officer (e.g., the supply officer) or an organizational staff officer (e.g., the S-4). The comptroller (or fiscal officer) acts as the principal financial advisor to the commander.

d. Communications and Information Systems

Communications and information systems collect, process, or exchange information. Under the cognizance of each element or subordinate organization G-6/S-6, these systems play an essential role in supporting C2 of the MAGTF. *See MCWP 3-40.3, MAGTF Communications System,* for a comprehensive discussion of this topic.

e. Billeting

Billeting provides safe and sanitary living quarters for assigned personnel and billet assignments are based on the operational circumstances. Commanders exercise their billeting responsibility through subordinate unit leaders. The commander's logistic officer (i.e. G-4/S-4) normally has staff cognizance of billeting facilities support. Billeting options include—

- Bachelor quarters in-garrison.
- Berthing compartments on ships.
- Tents in the field.

f. Food Service Support

Food service is a function of command. Commanders with a food service TO and TE provide food service support designated by the unit mission statement. Organizational food service responsibilities include—

- Accounting for all subsistence received.
- Properly storing all semi-perishable and perishable supplies.
- Adhering to sanitation practices.

- Preparing quality packaged operational rations (PORs), unitized group rations, (UGRs), and A-ration enhancements.
- Accounting for the number of personnel fed.
- Filing reports.

The LCE is responsible for—
- Providing support to the MEF beyond organic capability.
- Supplying and sustaining Class I (subsistence).
- Requisitioning of Class I (subsistence).

(1) Coordination of Resources. The selection of food service resources is METT-T dependent. The family of rations was developed for any situation. Detailed food service resources planning is conducted at the G-4/S-4 level in close coordination with food service officers, commanders, unit mess chiefs, and the LCE.

(2) Personnel Requirements. Food service personnel requirements are based on the feed plan, equipment, location, and numbers of static or remote feeding sites.

(a) Food Service Personnel. The required number of food service specialists to support mission-specific taskings should be determined by a subject matter expert (food service officer/food technician). The actual number of personnel will depend on the feed plan, equipment, location, and numbers of static or remote feeding sites.

(b) Messmen Support. Field mess attendants support is developed in coordination with the types of equipment and rations being used to sustain the operation. Specific requirements will be commensurate with the level of food service expected from the command. A traditional ratio of 1 mess attendant per 25 Marines being supported has been used for large scale field feeding operations. The MOS 3381 will not be assigned meal verification, cash collections, or mess attendant duties.

A messman (food service attendant) ratio of 1:18 (1 messman for every 18 troops embarked) will be used to compensate for the reduction of total food service specialists. The number of messmen furnished for the chief petty officer/staff noncommissioned officers mess will remain at a ratio of 1:15, and the number of wardroom messmen assigned will remain equal to 12 percent of the embarked officer population.

(3) Shipboard Feeding. Shipboard feeding must consider the following:

- Ratio of cooks for enlisted/staff noncommissioned officer (E1-E6) is: 1 cook per 72 embarked personnel, and for messmen the ratio is 1 messmen per 18 embarked personnel.
- At least 1 cook must be a staff noncommissioned officer (E6 or above). Sergeants/corporals (E5/E4) may be substituted for the remainder.
- With 1,000 embarked troops, one must be a sergeant (E5).

(4) Operational Rations. The following operational rations are to be used to feed Marines in the field. Safety, sanitation, and security protocols dictate utilizing approved sources of supply,

helps eliminate the risk of food tampering, and ensures that food vendors have security measures in place to protect the integrity of the supply chain down to the using unit.

(a) Individual Rations. The individual's rations are as follows:

- Meal, ready-to-eat (MRE) is a self-contained, individual field ration in lightweight packaging bought by the US military for its service members for use in combat or other field conditions where organized food facilities are not available.
- Meal, cold weather (MCW) and food packet, long range patrol (LRP) provides an operational ration for two separate operational scenarios, and requires hot water. The MCW is intended for cold-weather feeding; it will not freeze and supplies extra drink mixes for countering dehydration during cold-weather activities. The food packet, LRP is a restricted-calorie ration meant for special operations, where resupply is not available and weight and volume are critical factors.
- First strike ration (FSR) is a compact, eat-on-the-move assault ration designed for short durations of highly mobile, high-intensity combat operations. The intended purpose of the FSR is for usage during the first 72 hours of a conflict in lieu of using MREs.
- Meal, Religious, Kosher for Passover are rations that are utilized to feed those individuals in the military service who maintain a Kosher diet for Passover by providing three meals per day for not more than eight days during their observance of Passover.
- Meal, Religious, Kosher/Halal are rations that are utilized to feed those individuals in the military service who maintain a strict religious diet. Each meal consists of one Kosher or Halal-certified entree and religiously certified/acceptable complementary items sufficient to provide the recommended daily nutritional requirements.

(b) Group Rations. Group rations are as follows:

- Unitized group ration-heat and serve (UGR-H&S) is used to sustain military personnel during worldwide operations while at organized food service facilities. The UGR-H&S module is characterized by tray-pack entrees and starches/desserts.
- Unitized group ration-A (UGR-A) is used to sustain military personnel during worldwide operations while at organized food service facilities. The UGR-A includes perishable/frozen type entrees (A-Rations), along with commercial-type components. HQMC approval is required prior to requisitioning.
- Unitized group ration-B (UGR-B) and unitized group ration-M (UGR-M) are used primarily by the Marine Corps. They are designed to meet requirements for providing Marines with high-quality group rations that do not require refrigeration and are quick and easy to prepare.
- Ultra-high temperature-milk (UHT) is used by the Armed Forces as a mandatory supplement and/or enhancement for operational ration feeding during operations that do not have refrigeration capability or very limited capability.
- Modular operational ration enhancement (MORE) components are calorically dense and carry a balance of carbohydrates, caffeine, electrolytes, vitamins, antioxidants, and amino acids. All components can be eaten on the move without preparation and

are easy to consume and digest. The MORE is not intended to replace any individual ration under any circumstances. Rather, it is intended to be used by the warfighter in addition to their daily operational ration to provide the extra calories they need in high-stress, extreme environmental scenarios.

(c) Nutritional Guidelines/Consumption Parameters. Nutritional guidelines and consumption parameters are as follows:

- Operational rations consist of unitized group rations and individual rations (including individual restricted rations). These rations are designed for military personnel in a wide variety of operations and climates. The nutritional standards for operational rations (NSORs) are based on the military dietary reference intakes and are designed to support the special nutritional requirements for various expeditionary feeding situations.
- Unitized group rations menus are designed so the menus when used sequentially (e.g. day 1, day 2, day 3.) will meet the NSOR. The calculated or assayed nutrient content of edible portions of food as offered for consumption is compared to the NSOR. Total calories from fat will not exceed 35 percent of calories for these rations.
- Individual rations will not be consumed as the sole operational ration for more than 21 days. After 21 days, unitized group rations will be included in the daily mix of rations. This policy is based on extensive biochemical evaluations of consuming MREs for 30 days during field training. No degradation of performance or nutritional deficit was found before 21 days. When individual rations are the sole ration, units will request supplements and enhancements (for example, bread, milk, and fresh fruit and vegetables) when the logistical and tactical situation permits.
- The NSOR for individual rations do not apply to restricted rations. Restricted rations are nutritionally incomplete rations used in certain operational scenarios, such as the long-range patrol and reconnaissance, when troops are required to subsist for short periods carrying minimal weight. Restricted rations will not be consumed for more than 10 consecutive days.

(5) Operational Planning. Operational planning shall be based on the following ration mix:

- Days 1 to 21 consist of PORs (FSR or MRE).
- Days 22 to 90 consist of the following computation of the total personnel to be fed:
    20%—PORs, three meals per day.
    30%—UGR-H&S, two meals per day; and POR, one meal per day.
    50%—UGR-B (or by exception UGR-A), two meals per day; and POR, one meal per day.

Unitized ration meals are introduced into the feed plan as soon as the tactical, operational, and logistical situation permits; furthermore, "A" ration enhancements (fresh fruits, vegetables, bread, dairy, etc.) are also implemented as soon as the situation permits in order to provide a more nutritionally balanced meal. The feed plan is a standard form supplied by the food service officer to plan for the introduction of standard ration mixes in accordance with the unit mission and is METT-T dependent. For further information, reference MCRP 4-11.8A, *Marine Corps Field Feeding Program.*

(6) Field Feeding. In combat operations, expeditionary field messes are normally established at the battalion/squadron level in accordance with the Marine Corps Field Feeding Plan. The MEF, GCE, LCE, and ACE food service officers provide recommended sites, determine sizes of the facilities, designate which units to support and determine Class I support/sustainment for—

- Tactical feeding.
- Forward unit feeding:
  - Tray ration heater system (up to 250 personnel).
  - Enhanced tray ration heater system (up to 350 personnel).
  - Expeditionary field kitchen (up to 700 personnel).
- Base camp feeding:
  - Expeditionary field kitchen (500-700 personnel).
  - HNS (HHQ I&L approval).
  - Contract feeding (HHQ I&L approval).

g. Band

Traditionally, band members are trained in combat arms and may be used in a variety of roles, such as augmenting the headquarters defense in a combat environment. Designated major commands employ a military band to—

- Render honors.
- Provide military pomp at ceremonies.
- Perform on other occasions to raise or sustain morale.

h. Morale, Welfare, and Recreation

Activities, such as movies, special live-entertainment shows, and unit-level parties are MCCS opportunities used to relieve the stress and tedium of military operations. The MCCS is managed through command channels, with access to funds that support Marines at the unit level. Although MCCS activities are desirable, they should not interfere with mission accomplishment.

Chapter 3
Command and Control

"Command and control (C2) is the means by which a commander recognizes what needs to be done and sees that appropriate actions are taken" (MCDP 6, *Command and Control*). Through effective tactical-level logistic C2, commanders recognize and prioritize critical logistic requirements and direct the appropriate logistical and CSS response. This chapter describes procedures, responsibilities, and systems that are the means for executing tactical logistic and CSS C2 in the MAGTF. Command and control processes assist commanders in dealing with the following influences on warfare:

- **Uncertainty.** Commanders seek to clearly identify support requirements for tactical-level logistic and CSS operations. Absolute certainty will never be achieved in the dynamic situations characteristic of warfare. Commanders reduce uncertainty by employing a fully integrated planning process, prioritizing requirements, and ensuring redundancy and flexibility in their plans, as well as maintaining situational awareness.
- **Time.** There will rarely be enough time available to complete all desired planning and preparation for logistic operations, especially at the tactical level. Therefore, the assessment, planning, and execution cycle must be utilized to function effectively in the time available. This cycle is facilitated by a continuous exchange of information between all command echelons, functional activities, and exchange of liaison officers. See figure 3-1, on page 3-2, illustrates the assessment, planning, and execution conceptual process.
- **Tempo.** It is essential to maintain a constant, uninterrupted operational rhythm that leaves insufficient time for the enemy to react. To assist in maintaining a command's operational tempo, logisticians must anticipate support required and balance this with other battlespace activities. For example, attacks should not be interrupted or delayed because units need resupply or because LCE convoys are using critical main supply routes. To maximize operational tempo in this way, logisticians must participate fully in the operations planning process, stay updated on the status of battlespace activities, and prepare to conduct support operations.

The C2 for tactical-level logistics is focused on monitoring, directing, and executing logistic operations in support of tactical operations. Tactical logisticians establish and maintain communications links to higher, adjacent, and supporting and/or supported commands to ensure MAGTF elements can pass logistic information. See figure 3-2 on page 3-3.

3001. Establishing Command and Control

The MAGTF commander exercises C2 over MAGTF logistics. The commander evaluates logistic requirements based on subordinate organizations' capabilities, mission, and concept of operations. Based on this logistic evaluation, the MAGTF commander provides guidance to subordinate commanders. Typically, the guidance addresses three primary areas: requirements, priorities, and allocations. The subordinate commanders employ organic logistic resources to

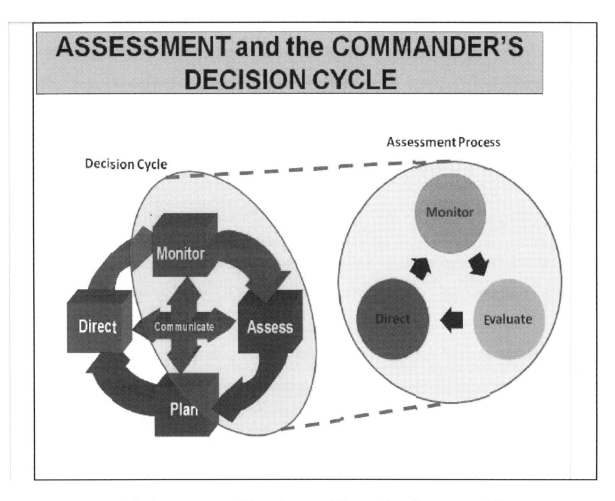

Figure 3-1. Assessment Planning and Execution Conceptual Process.

support their respective elements and then identify requirements beyond their organic capabilities to the LCE.

The LCE commander assigns support missions to subordinate elements based on the tactical situation, the supported unit's needs, and LCE capabilities. The LCE commander coordinates mission assignments with the MAGTF commander and supported unit commanders.

a. Command and Support Relationships

Inherent in command and support relationships is a clear understanding of the roles of each commander. The establishing commander, typically the MAGTF commander, will define the supporting to supported relationships, the degree of authority the supported commander has, and the overall priorities.

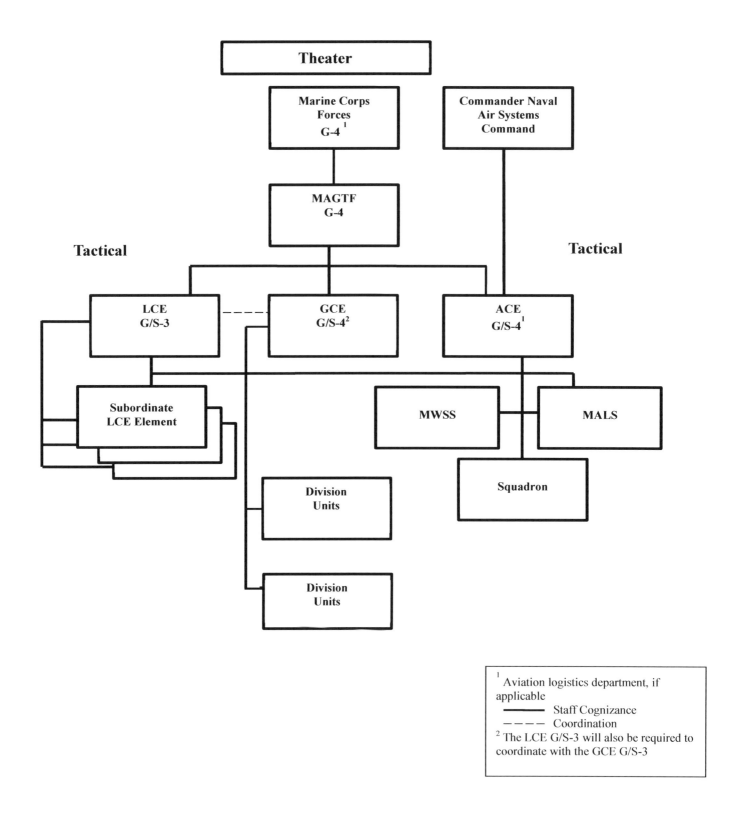

Figure 3-2. Staff Cognizance of Tactical-Level Logistics.

In general, the supported commander identifies his support requirements in terms of priority, location, timing, and duration. The supporting commander determines the forces, methods, and procedures to be employed in providing the support. If the supporting commander, subject to his existing capabilities and other assigned tasks, cannot fulfill the supported commander's requirements, then the establishing commander is responsible for determining a solution; i.e., a change in overall priorities or allocation of resources.

b. Command Relationships

Command relationships define higher and subordinate relationships between unit commanders. By specifying a chain of command, command relationships unify effort and enable commanders to use subordinate forces with maximum flexibility. Command relationships identify the degree of control of the gaining commander. The type of command relationship often relates to the expected longevity of the relationship between the headquarters involved and quickly identifies the degree of support that the gaining and losing commanders provide. Command relationships include organic, attached, administrative control, and direct liaison authorized (commonly referred to as DIRLAUTH):

- **Organic:** Assigned to and forming an essential part of military organization. The Marine Corps establishes command relationships through organizational documents such as table of organization and equipment (TOE) or through the means of Force Structure Review Group.
- **Attached:** Bound temporarily to a command other than its organic command. When attached, the unit is under the command of the unit to which it is attached. Unless otherwise stated, this encompasses all command responsibilities. Once the mission or function is completed, the attached unit returns to its parent unit.
- **Administrative control:** The exercise of authority over subordinate or other organizations in respect to administration and support, including organization of Service forces, control of resources and equipment, personnel management, unit logistics, individual and unit training, readiness, mobilization, demobilization, and discipline.
- **Direct liaison authorized:** The authority granted by a commander to a subordinate to directly consult, or coordinate an action with, a command or agency within or outside of the granting command.

These command relationships are established within the Marine Corps. When operating or participating in a joint environment, joint command relationships will be established by the commander of the joint forces. Command relationships used within the joint environment are combatant command (command authority), operational control (OPCON), tactical control (TACON), and support. For further information regarding joint command relationships refer to JP 3-0, *Joint Operations*. See table 3-1 on page 3-5.

Table 3-1. Command Relationship Matrix.

	Command Relationships							
	Then the inherent responsibilities are—							
If the relationship is—	Have a command relationship with—	May be task-organized by—	Unless modified, ADCON responsibility goes through—	Are assigned position or AO by—	Provide liaison to—	Establish/maintain communications with—	Have priorities established by—	Can impose on gaining command or support relationship of—
Organic	All forces organized with the HQ	Organic HQ	Higher HQ specified in organizing document	Organic HQ	N/A	N/A	Organic HQ	Attached; OPCON; TACON; GS;GSR;R;DS
Assigned	Combatant command	Gaining HQ	Gaining HQ	OPCON Chain of command	As required by OPCON	As required by OPCON	ASCC or Service assigned HQ	As Required by OPCON HQ
Attached	Gaining unit	Gaining unit	Gaining HQ	Gaining unit	As required by OPCON	As required by OPCON	Gaining unit	As required by OPCON
OPCON	Gaining unit	Parent unit and gaining unit; gaining unit may pass OPCON to lower HQ	Parent unit	Gaining unit	As required by gaining unit	As required by gaining unit and parent unit	Gaining unit	OPCON; TACON; GS; GSR; R; DS
TACON	Gaining unit	Parent unit	Parent unit	Gaining unit	As required by gaining unit	As required by gaining unit and parent unit	Gaining unit	OPCON; TACON; GS; GSR; R; DS

c. Support Relationships

Support relationships are direct support, general support, reinforcing, and general support-reinforcing (GS-R), as outlined in table 3-2 on page 3-6. Support relationships are not command authorities and are more specific than joint support relationships. Commanders establish support relationships when subordination of one unit to another is inappropriate. Commanders assign a support relationship when one of the following occurs:

- The support is more effective if a commander with the requisite technical and tactical expertise controls the supporting unit rather than the supported commander. The echelon of the supporting unit is the same as or higher than that of the supported unit. For example, the supporting unit may be a brigade, and the supported unit may be a battalion. It would be inappropriate for the brigade to be subordinated to the battalion; hence, the echelon uses a support relationship.

- The supporting unit supports several units simultaneously. The requirement to set support priorities to allocate resources to supported units exists. Assigning support relationships is one aspect of command.
- Support relationships allow supporting commanders to employ their units' capabilities to achieve results required by supported commanders. Support relationships are proceeded from an exclusive supported and supporting relationship between two units—as in direct support—to a broad level of support extended to all units under the control of the HHQ—as in general support. Support relationships do not alter administrative control. Commanders specify and change support relationships through task organization.

Direct support is a support relationship requiring a unit to support another specific unit and authorizing it to answer directly to the supported unit's request for assistance. (Joint doctrine considers direct support as a mission rather than a support relationship). A unit assigned a direct support relationship retains its command relationship with its parent unit, but is positioned by and has priorities of support established by the supported unit. A direct support mission requires a supporting unit to furnish close and continuous support to a single supported unit. Units given a tactical mission of direct support are not attached or under the command of the supported unit. An LCE unit that is in direct support of another unit is immediately responsive to the needs of the supported unit.

Table 3-2. Support Relationships Matrix.

	SUPPORT RELATIONSHIPS							
	Then the inherent responsibilities are—							
If the relationship is—:	Have command relationships with—	May be task-organized by—	Receives sustainment from—	Are assigned position or area of operations by—	Provide liaison to—	Establish and maintain communications with—	Have priorities established by—	Can impose gaining unit further command or support relationships by—
Direct Support	Parent unit	Parent unit	Parent unit	Supported unit	Supported unit	Parent unit: supported unit	Supported unit	See note 1
Reinforcing	Parent unit	Parent unit	Parent unit	Reinforced unit	Reinforced unit	Reinforced unit; parent unit	Reinforced unit: then parent unit	N/A
General Support-Reinforcing	Parent unit	Parent unit	Parent unit	Parent unit	Reinforced unit and as required by parent unit	Reinforced unit and as required by parent unit	Parent unit; then reinforced unit	N/A
General Support	Parent unit	Parent unit	Parent unit	Parent unit	As required by parent unit	As required by parent unit	Parent unit	N/A

Note: Commanders of units receiving direct support may direct a support relationship between their subordinate units and elements of the supporting unit after coordination with the supporting unit commander.

General support is that support which is given to the supported unit as a whole and not to any particular subdivision thereof. Units assigned a general support relationship are positioned and have priorities established by their parent unit. Parent commander retains complete authority over, and responsibility for the operation of the supporting unit. An LCE unit that is in general support supports the MAGTF under the direction of the LCE commander.

Reinforcing is a support relationship requiring a unit or force to support another supporting unit. Only like units (for example, artillery to artillery) can be given a reinforcing mission. A unit assigned a reinforcing support relationship retains its command relationship with its parent unit, but is positioned by the reinforced unit. A unit that is reinforcing has priorities of support established by the reinforced unit, then the parent unit.

General support-reinforcing is a support relationship assigned to a unit to support the force as a whole and to reinforce another similar-type unit. A unit assigned a GS-R support relationship is positioned and has priorities established by its parent unit and secondly by the reinforced unit.

d. Logistics Combat Element Command Relationships

Units of an LCE provide support to the other elements of the MAGTF, via either a general or direct support relationship. In a support relationship, the LCE unit, while responsive to the needs of the supported unit, remains under the command of its parent organization. The LCE commander retains control over subordinate units, which enhances centralized C2 and decentralized execution. While this is the normal method, it is not the only method. Both permanent and task-organized LCE units can be attached to other organizations. The MAGTF commander may direct the LCE commander to attach subordinate units to GCE or ACE units. The LCE commander retains responsibility for supporting subordinate units attached to other units but cannot assign or change their mission. See figure 3-3 on page 3-8.

e. Mission Assignments

A primary means of maintaining C2 over logistic units is the assignment of formal missions, particularly when LCE units function in a support relationship. The formalized mission structure helps by standardizing the responsibilities associated with each mission and allows the commander to tailor logistics to the tactical situation.

3002. Logistic and Combat Service Support Missions

Formal missions may be either standard or nonstandard. Standard missions are direct support and general support. A nonstandard mission is any remaining mission. Formal missions dictate relationships, responsibilities, and C2 procedures. They facilitate planning for future operations by providing for on-order tasks. They also simplify the planning and execution of MAGTF operations.

a. Inherent Responsibilities

Formal missions dictate specific responsibilities for both the supporting unit and the supported unit. Mission assignments establish the LCE unit's relationship to the supported unit, as well as to other subordinate units.

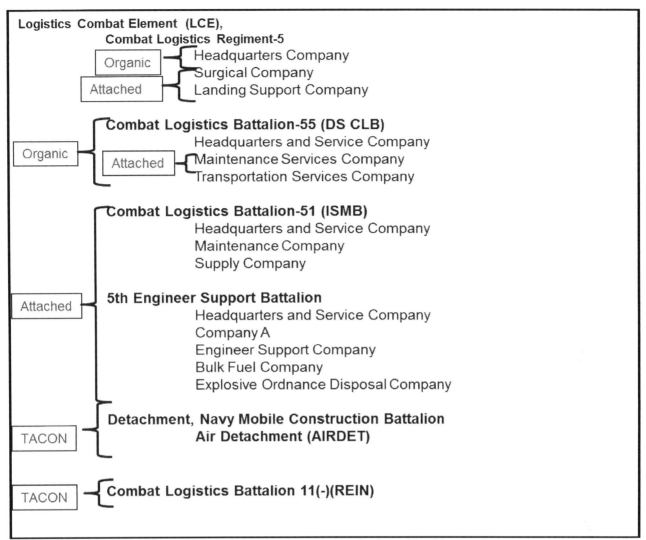

Figure 3-3. LCE Command Relationship Example.

An LCE unit or organization with a direct support mission responds to LCE request in priority from—

- Supported unit.
- LCE HHQ.
- Organic subordinate units.
- Provides liaison personnel to supported unit.

An LCE unit or organization with a general support mission responds to the LCE request in priority from—

- CSS HHQ.
- Supported unit.
- Organic subordinate units.
- Establishes liaison with supported units.

(1) Priority of Response. For each mission, the priority of response tells the supporting commander precisely who has priority of services. Support priorities are the primary distinction between standard missions.

(2) Liaison. Liaison is "that contact or intercommunication maintained between elements of military forces or other agencies to ensure mutual understanding and unity of purpose and action." *(JP 1-02)* The supporting commander decides what types of liaison to use. See paragraph 3007.

(3) Communications. Communications between the supporting and supported units is essential. The supporting commander, in conjunction with the parent HQ, decides what type of communications to use.

(4) Positioning. Positioning is not simply locating facilities on the ground. It includes the authority to displace facilities to new locations. The LCE commander has the responsibility and authority for determining the general location and the displacement time of ground-common subordinate units and facilities to ensure continued support to the MAGTF. The subordinate LCE commander recommends the time for displacements and selects exact locations for new facilities when given their general locale. The LCE units are often in areas that are under the control of other MAGTF elements, and, as such, the LCE commander must coordinate with those elements and the MAGTF commander before establishing or moving units and facilities.

b. Mission Statement Elements

Every LCE mission statement has four elements—three mandatory elements and one optional—identify the supporting unit, designate the standard mission assigned, and identify the supported unit. "CLB-1 conducts operations in direct support of 1st Marines" is an example of a simplified mission statement containing only the three mandatory elements.

The fourth element is optional and may be used to provide additional information and guidance. If the commander anticipates a change in mission, a fourth element may be added to the mission statement to facilitate future operations.

An example of an optional mission statement would be: 7th ESB (-) conducts operations in general support to I MEF. Attach one reinforced platoon to CLB 1, and one platoon in direct support of MWSS-372. Notice the LCE commander does not select the specific subordinate elements for alternative missions. Selecting specific platoons is the prerogative of the battalion and company commanders. It is, however, within the LCE commander's authority to direct different missions or command relationships for subordinate elements of the LCE and to task-organize subordinate elements. The LCE commander does so in coordination with the MAGTF commander, the supported unit commander, and the LCE subordinate commanders.

c. Standard Missions

An LCE unit assigned a direct support mission is immediately responsive to the needs of the supported unit. It furnishes continuous support to that unit and coordinates its operations to complement the concept of operations of the supported unit. The direct support mission creates a one-to-one relationship between supporting and supported units. The higher HQ of the supporting and supported units becomes involved only on a by exception basis. The supported unit sends requests directly to the supporting unit.

A direct support mission may be assigned to either a functional or task-organized LCE unit. A functional unit or a task-organized unit may be either a single function unit or a multifunction unit (provides support in two or more LCE functional areas). The following are examples of direct support missions assigned to functional and task-organized units:

- **Functional Units.** The LCE commander may assign the direct support mission to any functional subordinate organization (e.g., engineer or motor transport organizations).
- **Task-Organized Units.** The LCE commander may assign the direct support mission to a task-organized unit, such as a CLB, which is most often direct support. The commander must ensure that the task-organized unit has enough assets to accomplish the mission. Of particular concern is the ability to establish and maintain communications with the supported unit.
- **General Support.** A LCE unit assigned a general support mission supports the MAGTF or several units within the MAGTF under the direction of the LCE commander.

The general support mission is the most centralized mission. LCE commanders retain full control over their subordinate units, including establishing the priority of the units' efforts. This does not prevent supported units from dealing directly with various logistic agencies. For example, they submit requisitions directly to the supply source. However, the LCE commander may control how and when requisitions are filled. The LCE commander follows the priorities and allocations of the MAGTF commander. In certain cases, the MAGTF commander may stop the issue of supplies or items of equipment without prior approval of the LCE commander. In other cases, the MAGTF commander might specify a priority of issue for certain items or may assign a specific quantity to each unit.

The MAGTF LCE always has a general-support mission. However, the LCE commander may assign different missions to subordinate units consistent with the requirements of the tactical situation. The concept of logistics and CSS, found in Annex D and Annex W of the MAGTF OPORD, specifically addresses this topic and tells precisely how to satisfy the requirements of a particular tactical situation. The following are examples of general support missions assigned to functional and task-organized units:

- **Functional Units.** The LCE commander may assign the general support mission to any subordinate functional organizations. For example, the MLG commander may give the ESB the mission of general support of the MAGTF. The battalion would provide support based on the priorities of the MAGTF commander. The LCE commander would not assign this mission without prior coordination and approval from the MAGTF commander.
- **Multi-Functional Units.** The LCE commander may assign the general support mission to a task-organized unit such as a CLB or LFSP. Multi-functional logistic units may have sufficient assets to perform the functions associated with the mission. Of particular concern is the ability to establish and maintain communications and liaison with the supported unit and parent organization.

d. Nonstandard Missions

The LCE commander normally uses the direct support and/or general support standard missions to meet the needs of the supported force. However, unique situations may dictate the selection of a nonstandard mission. The nonstandard mission must satisfy the requirements of the specific situation and requires detailed planning and coordination.

The optional fourth element of the mission statement is the operative element in the nonstandard mission. The optional element amplifies the basic mission statement and addresses unique responsibilities and relationships.

The mission statement for a nonstandard mission must contain the three mandatory elements. For example, CLB-1 conducts operations in general support of assigned US and multinational forces. The optional fourth element, which gives advance information on subsequent missions, may also be used, as appropriate.

The mission statement above is adequate for a nonstandard mission. For the CLB-1 commander, however, it does not provide enough information in this particular case. With standard missions, the CLB commander immediately knows the associated responsibilities. When assigning a nonstandard mission, the LCE commander must also give detailed coordinating instructions to amplify the mission statement. Paragraph 3 of the LCE OPORD should include priority of response to support requests, liaison requirements, and communication requirements.

Priority of response to support request for—

- MAGTF units (or name of specific unit).
- Other US forces.
- Allied forces (classes I, III, and V only).

Liaison requirements—

- Maintain liaison with supported Marine Corps units on a full-time basis.
- Maintain liaison with other supported units, as required.

Communications responsibilities—

- Establish and maintain communications with MAGTF units on a full-time basis.
- Establish and maintain communications with other elements, as required (general support/direct support defined).

3003. SUPPORT PROCEDURES IN TACTICAL LOGISTIC FUNCTIONAL AREAS

The functional areas of tactical-level logistics are managed with procedures tailored to support particular functions.

a. Supply

MAGTF commanders delegate to LCE commanders the responsibility of managing the flow of support from source to consumer. Three management techniques and procedures are critical to supply support.

(1) Control. Supplies should flow by the most direct route from the source to the consumer. The LCE units should handle supplies as infrequently as possible.

(a) Records. Records should include only information that is essential to control supply activities and to ensure sustainability.

(b) Stockage Objective. The stockage objective is the maximum quantity of materiel that the LCE must have on-hand to sustain current operations. It consists of the sum of stocks represented by the operating level and the safety level. The operating level is the level required to sustain operations between submissions of requisitions or between the arrival of successive shipments. These quantities are based on the established replenishment period (daily, monthly, or quarterly). In combat, the replenishment period is usually shorter than during peacetime operations. The safety level is the quantity required to continue operations if there are minor delays in resupply or unpredictable changes in demand. In combat, the safety level is more critical than during peacetime.

The MAGTF commander prescribes the stockage objective for CSS installations on the basis of the recommendations of the LCE commander. Selection of the proper stockage objective is critical for proper management of transportation and continued support of combat operations. If the stockage objective is too high, it can place an excessive burden on handling and management systems. If the stockage objective is too low, it can delay or even prevent combat operations.

(c) Reorder Point. The reorder point is that point at which the CSS unit must submit a requisition to maintain the stockage objective. The supply representative requisitions the stockage objective when the sum of the requisition processing time, shipping time, and safety days of supply equals the remaining days of supply based on daily consumption rates. For example:

		Days of Supply
Safety level	=	5
Reorder time	=	2
Shipping time	=	15
Reorder point	=	22

(2) Distribution Methods. The two normal methods of distribution are supply point distribution and unit distribution, but the commander typically uses a combination of the two methods.

(a) Supply Point Distribution. In supply point distribution, the supported unit picks up the supplies from a central point established by the supporting unit similar to getting fuel from a filling station or food from a store.

(b) Unit Distribution. In unit distribution, the supporting unit (e.g., LCE) delivers supplies to the supported unit. The supported unit will, in turn, distribute the supplies to subordinate elements.

(c) Combination. Normally, the commander uses a combination of unit and supply point distribution. The commander assigns top priority for unit distribution to those units that are in contact with the enemy and that have limited organic transportation. The commander gives a lower priority to engaged units with more organic transportation. The lowest priority is assigned to units not in contact with the enemy. When the available transport has been allocated to unit distribution, the remaining support requirements must be satisfied through supply point distribution.

(3) Replenishment Systems. Replenishment systems are either pull systems, push systems, or a combination of both. Selecting a replenishment system is generally based on the availability of supplies and distribution capabilities.

(a) Pull Systems. A pull system requires the consumer to submit a support request. This system provides only what the supported unit requests. Pull systems generally do not anticipate a unit's needs, which make them less responsive, but more efficient than push systems.

(b) Push Systems. Push systems use reports as the requesting document or anticipate demand based on consumption rates. For example, on-hand and usage reports submitted by the supported unit serve as the basis for resupply. The LCE delivers sustainment based on consumption rates and the desired basic load of the unit without waiting for a requisition. Use of this method could burden the unit with more supplies than it can handle, which makes them more responsive but less efficient.

(c) Combination. The MAGTF commander should specify the most appropriate replenishment system, which is often a combination of the two methods. The decision should be based on the tactical situation, available resources, and the recommendations of the LCE commander.

b. Maintenance

The goal of maintenance support operations is to keep equipment operational at the using unit. Supporting commanders achieve this goal by balancing centralization of control with decentralization of execution. Maintenance support procedures need to be flexible and adaptable to changing situations. For example, during the amphibious assault, both the LFSP and supported organizations have limited maintenance capabilities. As a general rule, the goal in combat should be centralized control with decentralized execution to maximize responsiveness. Organizational contact teams, from the owning organizations and intermediate MSTs from the LCE go forward and repair equipment whenever possible.

c. Transportation

The MAGTF commander generally centralizes control of movement at the highest level. Movements should be regulated and coordinated to prevent congestion and conflicting movements over transportation routes. The transportation system must be highly adaptable to use the MAGTF's limited transportation capabilities effectively. This adaptability enables the commander to maintain continuous movement of personnel, supplies, and equipment.

Commanders must maximize the efficient and effective use of transportation assets. The commander must keep equipment loaded and moving, while allowing for adequate maintenance and personnel rest.

d. General Engineering

The LCE engineer staff officer coordinates execution of general engineering projects with the ESB, NCF commander (as necessary) and the MAGTF engineer staff officer. Within the ACE, the MWSS provides limited general engineer support to meet ACE unique requirements at FOBs or FARPs. Critical to the coordination of tactical level general engineer support is the prioritization and allocation of transportation and material resources required for construction projects (base camps, airfields, roads, obstacles/barriers, bridging). The MAGTF can request augmentation from Army, Navy or Air Force units. Additional information pertaining to general engineering capabilities is contained in MCWP 3-17.7 and MCWP 3-21.1. These provide description of MWSS capabilities and TTP. See MCWP 4-11.5, for additional information pertaining to NCF capabilities to support general engineering requests of the MAGTF.

e. Health Service Support

The medical regulating system is activated as necessary for monitoring and controlling the movement of patients through the casualty evacuation and HSS system. The medical regulating system is responsible for patient movement and tracking through successive levels of medical and dental care to provide the appropriate level of care. For information on medical regulating procedures, see MCRP 4-11.1G and JP 4-02.2.

f. Services

The services function provides for the effective administration, management, and employment of military organizations, as previously discussed. The administrative subfunctions are categorized as either command support or combat service support.

3004. COMMAND GROUPS AND CONTROL AGENCIES

Each MAGTF element establishes sections to direct operations and control employment of their organic, ground-common and aviation-peculiar logistic capabilities. Additionally, they will coordinate CSS requirements with the LCE.

a. Aviation Ground Support Operations Center

The MWSS will establish an aviation ground support operations center to control aviation ground support tasks at the ACE airfields. The center coordinates the activities of the airfield operations, motor transport operations, engineer operations, medical, and other services sections.

b. Ground Combat Element Logistic Operations Center

The GCE establishes a logistic operations center that controls and coordinates day-to-day operations within the GCE organization. The logistic operations center focuses on meeting the needs of the supported units. The GCE G-4/S-4 logistic officer supervises the functioning of the logistic operations center. Optionally, there are GCE units that combine their administration section with the logistic section to form an administration and logistic operations center (ALOC). This is not a doctrinal term but there is a possibility to incorporate the administration and logistics to provide additional benefits, dependent on the mission situation.

c. Logistics Combat Element Combat Operations Center

The LCE establishes a LCE combat operations center (COC) that controls and coordinates the day-to-day operations of the LCE and focuses on meeting the needs of supported units. The LCE operations officer supervises the day-to-day functioning of the LCE COC.

The LCE commander establishes the LCE COC in the LCE command post. The LCE COC continually monitors and records the status of logistic operations and its personnel coordinate and control CSS operations according to the established policies, SOPs, and operational decisions of the commander.

The LCE COC monitors the CSS request communication network. The LCE COC has direct lines to subordinates, supported units, and higher HQ, and it may have data links. Normal LCE COC functions include the following:

- Receiving and recording operational reports from subordinate units.
- Maintaining current plots of the friendly and enemy situation and displaying the information in the LCE COC.
- Preparing and submitting operational reports to HHQ.
- Providing dedicated communications channels for control of logistic operations.
- Transmitting orders and decisions.
- Monitoring the progress of ground-common logistic operations and reporting significant events and incidents to the commander.
- Monitoring and deconflicting route allocations for logistic convoy operations.
- Advising interested staff sections of events or information of immediate concern to them.
- Serving as the principal point of contact for liaison personnel from senior, supported, or adjacent units.
- Maintaining a rear area security overlay that depicts preplanned targets, active security measures for logistic installations, and main supply routes within the rear area.
- Coordinating security of logistic installations and main supply routes within the rear area with higher and adjacent elements of the MAGTF.

The LCE COC is not a separate organization. The LCE unit's operations and communications personnel staff the LCE COC. Local SOPs govern the size and composition of the LCE COC. Generally, the commander has the following LCE COC organization configuration options:

(1) Centralized Logistics Combat Element Combat Operations Center. Figure 3-4, on page 3-16, depicts a centralized LCE COC arrangement. An advantage to placing functional representatives for supply, maintenance, transportation, engineering, health services, and services within the LCE COC is that the watch officer has immediate access to technical advice. This option is appropriate when tactical considerations do not require dispersal. A disadvantage can be the high activity level generated by large numbers of personnel and communications in a confined facility. Higher-level LCE organizations and those farther to the rear use a centralized LCE COC more frequently than do smaller units.

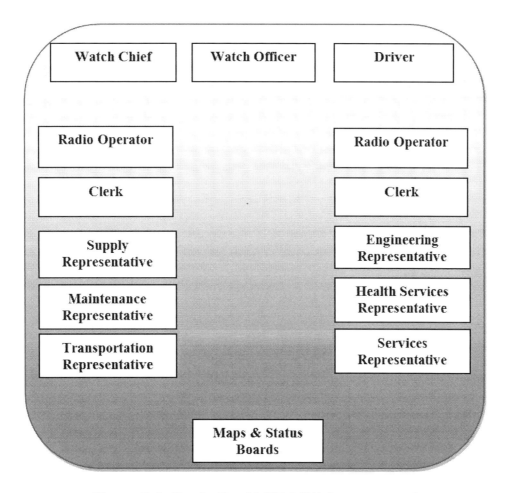

Figure 3-4. Centralized LCE COC Arrangement.

(2) Decentralized Logistics Combat Element Combat Operations Center. Figure 3-5, on page 3-17, depicts a decentralized LCE COC arrangement with functional representatives placed outside the LCE COC. Smaller LCE organizations and those farther forward most often select this option. In some situations, the LCE unit will not have enough personnel or skills to operate a centralized LCE COC. In other cases, dispersion is a tactical necessity that weighs against centralization.

3005. MOVEMENT CONTROL ORGANIZATIONS

Movement control combines the planning, routing, scheduling, and control of personnel and cargo movements over (LOC) to support the deployment of forces. This paragraph discusses movement control techniques and management agencies. When operating as part of a joint, allied, or coalition force, the MAGTF commander follows the distribution management and movement control regulations of that command. Normally, the higher commander establishes a movement control agency to provide movement management services and highway traffic regulation. This agency coordinates with allied and host nation movement control agencies.

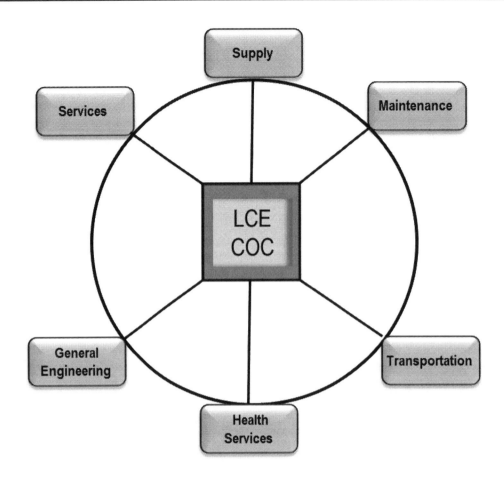

Figure 3-5. Decentralized LCE COC Arrangement.

a. Movement Control Centers

Movement control centers (MCCs) are agencies that plan, route, and schedule personnel, supplies, and equipment movements over LOCs (point of origin to POE, port of debarkation (POD) to final destination or movements within the AO).In some cases, the agencies are permanent. For example, every MAGTF should have a full-time distribution and transportation section. For smaller MAGTFs, this may be no more than one or two Marines in the combat service support operations center. In other cases, movement control agencies are temporary. Battalions, squadrons, regiments, and groups establish temporary movement control centers when their organizations are moving. Local standing operating procedures (SOPs) establish the composition and procedures for MCCs. Figure 3-6, on page 3-18, depicts the relationships between various commands, their movement control agencies, and supporting organizations during deployment and sustainment distribution of a MAGTF. Figure 3-7, on page 3-18, depicts movement control in garrison.

b. MAGTF Deployment and Distribution Operation Center

The Marine air-ground task force deployment and distribution operations center (MDDOC) is the MAGTF commander's agency responsible for the control and coordination of all deployment

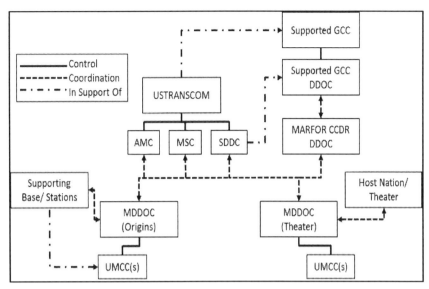

Figure 3-6. Movement Control Relationships During Deployment.

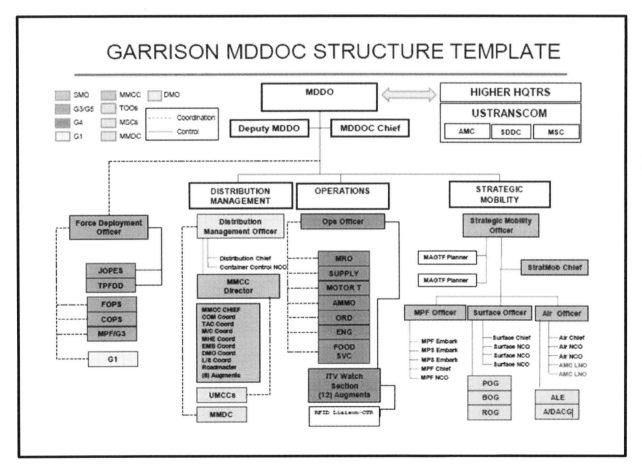

Figure 3-7. Movement Control Relationships in Garrison.

support activities. It is also the agency that coordinates with geographic combatant commander's's DDOC, and USTRANCOM's transportation component commands (e.g. AMC, SDDC, MSC). When the MAGTF operates as part of a joint force under a JFC, the MDDOC coordinates with USTRANSCOM via the JFC's joint deployment and distribution operations center (JDDOC), which coordinates requirements for all eographic combatant commander's service components. The MDDOC (or tailored capability), located within the MAGTF command element, conducts integrated planning, provides guidance and direction, and coordinates and monitors transportation resources in its role as the manager of the MAGTF's theater and tactical distribution processes. The MDDOC may operate under either the G-3 or G-4 but requires integration with each counterpart staff.

c. MAGTF Materiel Distribution Center

The MAGTF Materiel Distribution Center (MMDC) is the MAGTF's distribution element. The mission of the MMDC is to provide general shipping and receiving services consolidated distribution services and to maintain asset visibility to enhance throughput velocity and sustain operational tempo. The MMDC is located in the MLG for garrison operations. While in garrison, the MMDC will make every effort to integrate/collocate with the base freight operation, in order to maintain distribution competence. For deployed operations, the MMDC resides with the logistics combat element (LCE). The MMDC will establish and operate the distribution network in a deployed environment.

d. Distribution Liaison Cells

Distribution liaison cells are distribution elements that are manned by the LCE. Distribution liaison cells are task-organized and structured to perform various tasks at POEs/PODs aboard Marine expeditionary units (MEU's) or forward operating areas to include but not limited to providing support for deploying MAGTFs.

e. Terminal Operations Organizations

Terminal operations organizations are integral to the deployment and distribution system by providing support at strategic, operational, and tactical nodes. Terminal operations organizations are established under the operational control of the MMCC and/or the MDDOC. Examples of terminal operations organizations are AACG/DACG, port operations group, beach operations group, railhead operations group, and the movement control agency (MCA) of the landing force support party (LFSP). TOOs will be task-organized, manned, and augmented by MSCs, as required, to perform these tasks.

f. MAGTF Movement Control Center

The MMCC is a standing organization and the subordinate element of the MDDOC that allocates, schedules, and coordinates internal transportation requirements based on the MAGTF commander's priorities.

> Note: At the MEF level, the MMCC replaced the logistics movement control center.

The MMCC supports the planning and execution of MAGTF movements. The MMCC coordinates all MAGTF ground movement scheduling, equipment augmentation, transportation requirements, material handling equipment, and other movement support. In a theater of operation, the MMCC coordinates and deconflicts ground movements on theater controlled routes, and register requirements to the joint movement center (JMC) for support. In addition, the MMCC coordinates activities with installation operations, supporting commands, and MSC UMCCs.

g. Major Subordinate Command Unit Movement Control Center
Division, wing, and MLG commanders deploy forces to support operational MAGTFs. Both deploying and employing MSC commanders manage transportation and communications assets needed to execute deployments. Each command at the MSC level and below activates its UMCC to support marshaling and movement of assigned subordinate units. Major subordinate commands establish UMCCs down to the battalion, squadron, or independent company, as required, to serve as the unit transportation capacity manager (TCM).

h. Base Operations Support Group
Bases from which Marine Corps operatinf forces units deploy establish base operations support groups to coordinate their efforts with those of the deploying units. Bases operations support groups coordinate and manage transportation, communications, and other functional support requirements beyond organic capabilities to supported units during deployment similar to MARFOR component commands.

i. Station Operations Support Group
Air stations from which Marine Corps operating forces deploy establish station operations support groups to coordinate their efforts with those of the deploying units. Like major MARFOR commands, air stations have transportation, communications, and other assets useful to all commands during deployment.

j. Flight Ferry Control Center
In addition to its MCC, the aircraft wing establishes a flight ferry control center to control deploying aircraft. The flight ferry control center operates under the MAW G-3.

3006. MARITIME PREPOSITIONING FORCE ORGANIZATIONS

The arrival of the MPF and its assembly into a fighting force are critical operational concerns of the MEF in general and the MLG in particular. The MEF forms a number of temporary organizations whose purpose is to transform the cargo and personnel of an MPF into a viable combat force.

a. Survey, Liaison, and Reconnaissance Party
The survey, liaison, and reconnaissance party is a self-sustaining task organization formed from the MAGTF and Navy support element. It conducts reconnaissance, establishes liaison with in-theater authorities, and initiates preparations for the arrival of the main body of the fly-in echelon (FIE) and the MPSRON. The survey, liaison, and reconnaissance party normally deploys to the arrival and assembly area (AAA) under MAGTF cognizance.

b. Offload Preparation Party
The offload preparation party is a temporary task organization that consists of maintenance technicians, embarkation specialists, and equipment operators drawn from all elements of the

MAGTF and the Navy support element. It prepares equipment on board the MPSs for debarkation into the AAA. The offload preparation party can join the MPS before sailing, during transit, or on arrival into the AAA. Ideally, the offload preparation party boards the MPS 96 hours before arrival into the AAA.

c. Arrival and Assembly Operations Group

The AAOG is a temporary task organization that controls and coordinates the arrival and assembly operations of MPFs. Normally, the AAOG deploys as an element of the advance party and initiates operations at the arrival airfield. The AAOG is formed from elements of the MAGTF and liaison personnel from the Navy support element during an MPF operation. The AAOG controls the following four subordinate throughput organizations:

- The port operations group is responsible for preparing the port prior to arrival of the MPS and for the throughput of equipment and supplies as they are offloaded from the ships.
- The beach operations group organizes and develops the beach area as necessary to support the offload and throughput of equipment and supplies.
- The AACG is responsible for the control and coordination of the offload of airlifted units and equipment at the airfield.
- The MCC plans, schedules, routes, and controls the movement of personnel, equipment, and supplies from the port, beach, or airfields to the unit assembly areas.

3007. AMPHIBIOUS SHIP-TO-SHORE MOVEMENT ORGANIZATIONS

The Navy control organization provides positive centralized control of STS movement. Close coordination among the waterborne and helicopterborne movements and supporting, pre-landing and in-stride operations with the flexibility to change the landing plan is required. This coordination ensures maximum tactical effectiveness during the landing and subsequent buildup of infrastructure or combat power ashore. The LFSP lands with surfaceborne units to facilitate the flow of personnel, equipment, and supplies across the beach and beyond and to establish a beach support area to provide CSS to these units. Requests for on-call waves prepositioned emergency supplies, nonscheduled units, and adjustments to the landing plan are made by tactical commanders through the LFSP to the TACLOG group detachment for the required liaison with primary control officers to provide the tactical units, CSS, or adjust the landing plan. To facilitate required liaison between landing force units ashore and the Navy control organization, the TACLOG detachment at each LF echelon is embarked in the same ship with the Navy control organization exercising OPCON over the STS movement of that LF echelon.

a. Navy Control Organization

The Navy is responsible for control of the ship-to-shore movement of both waterborne and assault support forces. The structure of the Navy control organization varies depending on the scope of the operation and number and type of beaches and landing zones. The TACLOG group is the Marine Corps agency for advising and assisting the Navy control organization regarding landing force requirements during the STS movement.

(1) Control for Waterborne Movement. The following officers are responsible for controlling waterborne STS movements:

- **Central Control Officer.** Normally aboard the amphibious task force (ATF) flagship, the central control officer directs the movement of all scheduled waves. After scheduled waves have landed, the central control officer continues to coordinate movement to and from the beach until unloading is complete. The central control ship is normally some distance seaward of the line of departure.
- **Primary Control Officer.** The senior Navy commander appoints a primary control officer for each transport organization that lands a regimental landing team across a colored beach or a geographically separated beach. From aboard the primary control ship, this officer directs movement to and from a colored beach. The primary control ship is usually near the line of departure.

(2) Control for Assault Support Operations. The senior Navy commander, through the tactical air officer, controls helicopters during the STS movement. Control agencies include the Navy tactical air control center (TACC) and amphibious air traffic control center (commonly referred to as AATCC). These agencies control helicopters to meet both tactical and logistic requirements. They also coordinate the movement of helicopters with other aircraft movement. The helicopter control system must be flexible and responsive to the requirements of the tactical situation. See MCWP 3-31.5, *Ship-to-Shore Movement*, for more information.

b. Landing Force Control Organization

The landing force control organization interfaces with the Navy control organization to keep it apprised of landing force requirements and priorities as well as to advise on transportation methods and phasing of serials. Although the exact structure of the landing force control organization varies, it is usually composed of the landing force operations center (LFOC), LFSP, and TACLOG group.

(1) Landing Force Operations Center. During the initial phases of the amphibious operation, the LFOC is the MAGTF commander's command post afloat. Normally, the LFOC is located in the vicinity of the ATF's combat information center. The LFOC maintains communications with the LFSP elements and with the landing force TACLOG group, which functions as the landing force liaison with the Navy control organization through the central control officer. From the LFOC, the MAGTF commander—

- Monitors the progress of the STS movement and operations ashore.
- Controls assigned assault units.
- Communicates with subordinate commanders.

(2) Landing Force Support Party. The ship-to-shore movement is a complex evolution that generates intensive activity under combat conditions. The LFSP is a temporary landing force organization composed of Navy and landing force elements tasked to provide initial combat support and CSS to the landing force during the STS movement. Its mission is to support the landing and movement of troops, equipment, and supplies across the beaches and into landing zones. The LFSP facilitates the smooth execution of the landing plan. It is specifically task-organized to facilitate a rapid buildup of combat power ashore by ensuring an

organized and uniform flow of personnel, equipment, and supplies over the beach in support of the landing force scheme of maneuver.

(3) Tactical-Logistical Group. At the landing force level, the TACLOG group is composed of representatives from the MAGTF G-3/S-3 and G-4/S-4. The TACLOG group advises the Navy control organization of the STS movement requirements to meet the tactical requirements ashore and to assist in identifying support resources. To provide this advice, the TACLOG group—

- Keeps abreast of which serials have landed.
- Monitors the command, tactics, and logistical nets to anticipate requirements ashore for serials.
- Provides the central control officer with advice on the priority of landing additional serials.
- Recommends modes of transportation for serials, when appropriate.

The TACLOG group is subordinate to the landing force TACLOG group established by each subordinate commander in the landing force. A subordinate TACLOG group may also be established aboard the helicopter transport group commander's ship to provide liaison for the helicopterborne force. These subordinate TACLOG groups coordinate duties between the Navy control organization, the landing force, and the landing force TACLOG group.

c. Naval Beach Group

The NBG is "a permanently organized naval command within an amphibious force comprised of a commander and staff, a beachmasters unit, an amphibious construction battalion [ACB], and an assault craft unit, designed to provide an administrative group from which required naval tactical components may be made available to the attack force commander and to the amphibious landing force commander to support the landing of one division (reinforced)." (JP 1-02) This group task-organizes beach party teams and/or groups for specific tasks. It can make limited beach improvements to help in the landing and the evacuation of casualties and EPWs. For additional information on the NBG, refer to Navy, Tactics, Techniques, and Procedures (NTTP) 3-02.1M, *Ship-to-Shore Movement*, and JP 3-02.1 *Amphibious Embarkation and Debarkation*. The NBG is an administrative organization that provides—

- A beach party.
- Pontoon causeway teams.
- Self-propelled pontoon barges.
- Elements for lighterage or transfer line operations.
- Warping tug teams for tending causeways and salvage.
- STS bulk fuel elements.
- Underwater wire communications from the primary control ship to the beach.

d. Other Navy Landing Support Assets

(1) Navy Cargo-Handling Battalion. A mobile logistic support unit that is organized, trained, and equipped to load and off-load Navy and Marine Corps cargo carried in MPSs and

merchant break bulk or container ships in all environments. Navy cargo handling battalion also loads and off-loads Navy and Marine Corps cargo carried in military-controlled aircraft. Additionally it operates an associated expeditionary air cargo terminal.

(2) The Navy's Fleet Survey Team. Navy's fleet survey team may conduct quick response hydrographic surveys and produce chart products in the field to support maritime requirements. They provide high resolution hydrographic surveys for use in nautical or tactical charting, support amphibious landings, mine warfare, or naval special warfare with bathymetry and other collected hydrographic information. Deployable detachments from this team can conduct navigation quality surveys or clearance surveys to provide access to ports and waterways in support of amphibious operations. In forward-deployed amphibious readiness group/MEU operations in which access to the above teams may be more difficult, commanders should use organic or supporting reconnaissance and surveillance assets to provide hydrographic information.

3008. COMMUNICATIONS

Commanders must establish communications with higher, adjacent, and subordinate commands to promote situational awareness and to direct and coordinate military operations. Following the MAGTF communications plan, commanders establish single-purpose and general-purpose nets and/or frequencies for the control of MAGTF and/or element operations, logistic and CSS operations, and general administrative support.

The communications plan must specify the means for requesting and coordinating ground-common and aviation-peculiar logistic support. In addition, the plan must designate the primary and alternate means for logistic communications.

The logistics request communication network is the most important communication network for day-to-day ground-common logistic operations. The request network is a direct link between the CSS unit and the supported organizations. Each LCE organization establishes a logistic request communication network. Supported organizations enter the network to pass routine or emergency support requests. The LCE also establishes a CSS request network between itself and its subordinate LCE organizations and uses these communication networks to pass reports, requests, and orders. Supported unit logistic officers should monitor the logistic request network to assess the status of LCE units and to facilitate anticipation of requirements.

3009. LOGISTIC INFORMATION MANAGEMENT

Logistic C2 manages the process of providing resources to support the warfighter; information management is a principal tool in this process. Tactical-level logistic information management ranges from manual methods to employment of logistics automated information systems.

a. Organic Capabilities
Most Marine Corps organizations down to company levels have organic information systems capabilities to manage their personnel, logistic, and training responsibilities.

The communication and information systems officer (G-6/S-6) supervises the command's communications and information systems support operations. The G-6/S-6 is responsible for the technical direction, control and coordination of communications and information systems support tasks. The G-6/S-6 section works closely with the functional users of AIS to ensure its efficacy.

b. Information Systems Functional User Responsibilities

Functional users of information operate the information systems supporting their functional area. Functional users include every staff section that is supported by communications and information systems. Consequently, all staff principals have functional user responsibilities for the functional area over which they have staff cognizance. For example, the logistic staff officer has functional user responsibilities for the Global Combat Support System-Marine Corps (GCSS-MC). Functional user responsibilities include—

- Serving as the primary point of contact for issues affecting information systems supporting the functional area.
- Conducting the following routine information system administration:
    - Assigning user identification, passwords, and privileges.
    - Performing data and/or file storage and management.
    - Conducting system backups.
    - Coordinating with the G-6/S-6 to ensure that adequate hardware, software, trained personnel, and procedures are in place before implementing a new system or system modification.
    - Coordinating with the G-6/S-6 to develop and maintain user training programs for communications and information systems.
    - Identifying to the G-6/S-6 information system support requirements. Identifying to the G-6/S-6 specific communications requirements, including requirements to interface with other information systems and potential interface problems.
    - Complying with applicable communications and information systems security measures. Reporting malfunctions and outages and coordinating with the G-6/S-6 to restore service.
    - Designating an information management officer for the staff section.

c. Information Systems

Each MAGTF element has computers and required software to support data input for standard logistic information systems, as well as to operate related C2 systems. These logistic information systems include manpower, supply, maintenance, transportation, embarkation, disbursing, contracting and aviation-peculiar systems.

(1) Global Command and Control System. The Global Command and Control System (GCCS) is a suite of software applications and hardware designed for planning, execution, C2 of forces, and multidiscipline intelligence processing. The system receives multiple sensor feeds and reports that assist in the development of the common operation picture. Planning and execution applications also support supply, maintenance, transportation, acquisition, finance, personnel, engineering, and force health protection. As a system, it supports the Joint Chiefs of Staff and combatant commanders through the Joint Operation Planning and

Execution System (JOPES) in a contingency and crisis planning. For more information see MCWP 3-40.3.

The JOPES is a DOD system of joint policies, processes, procedures, and reporting structures, supported by communications and information technology that is used by the joint planning and execution community to monitor, plan, and execute mobilization, deployment, employment, sustainment, redeployment, and demobilization activities associated with joint operations.

(2) Global Status of Resources and Training System. The Global Status of Resources and Training System (GSORTS) is an automated DOD system used to provide the Secretary of Defense, Chairman of the Joint Chiefs of Staff, and other senior DOD officials with authoritative identification, location, and resource information on DOD units/organizations. The GSORTS provides monitoring information to the National Military Command System. The Chairman of the Joint Chiefs of Staff Instruction (CJCSI) 3401.02 series and the Chairman of the Joint Chiefs of Staff Manual (CJCSM) 3150.02 series provide guidance to all service headquarters regarding requirements for reporting the status of resources and training in its units to GSORTS. The Marine Corps feeds into GSORTS via the Defense Readiness Reporting System (commonly referred to as DRRS).

(3) Global Combat Support System-Marine Corps. The GCSS-MC aims to maximize Marine Corps combat effectiveness through logistic information technology. The GCSS-MC will enable an end-to-end logistic chain that is agile, responsive, flexible and reliable. This system provides improved processes, driving quantifiable changes for precision distribution and logistic chain management. Additionally, it provides cross-functional information to enhance in-transit visibility and total asset visibility, thus affording timely decisions for logistic chain management through the last mile. Program benefits also include a reduction in customer wait time, a decreased dependency on forward positioned materials, and less frequent redundant requisitioning. The system controls inventory issues and will allow Marines to adjust on-hand inventories downward, increase inventory accuracy and validity, and improve initial inventory fills. The GCSS-MC will modernize, integrate, and sustain information technology solutions for the Marine Corps logistic community, providing the right logistic data, at the right time, and right place. The end state will be a successfully implemented information technology system utilized by the MAGTF and supporting establishments to enhance their logistic warfighting capability with minimal disruption to the enterprise network.

(4) Unit Readiness Planning/Unit Training Management. Unit readiness planning and unit training management focus training on the tasks that are essential to a unit's wartime capabilities. Unit training management is the use of the Systems Approach to Training and the Marine Corps Training Principles in a manner that maximizes training results and focuses the training priorities of the unit in preparation for the conduct of its wartime mission. Marine Corps units update and maintain Marine Corps Training Information Management System to schedule and track training and exercises.

The suite of logistic automated systems is the primary tool for managing the support capabilities of a MAGTF. These systems support ground common logistic data requirements that provide updates to JOPES to support force deployment, planning, and execution and are as follows:

- Marine Air-Ground Task Force Deployment Support System II (commonly referred to as MDSS II) enables commanders at various echelons of a MAGTF to build and maintain a data base containing force and equipment data that reflects how the MAGTF is configured for deployment. The Data can be maintained during normal day-to-day garrison activities and updated during plan development and execution.
- Transportation Coordinator's Automated Information for Movement System II (TC-AIMS II) is designed to enable users to manage all aspects of transportation operations. The TC-AIMS II provides automated support to functions performed by a wide range of users including unit movement officers, installation transportation officers, and mode managers responsible for transportation and distribution in support of the full continuum of operations. The TC-AIMS II includes automated support to assist unit commanders to create, maintain, manage, and update unit equipment, personnel, and deployment information. It facilitates planning and execution of organic movements incorporating the mechanism for identifying assets and requirements for force deployment/redeployment on deliberate and crisis action planning.
- The GCSS-MC is another primary logistics automated information system. It provides automated support for ground-common supply and maintenance.
- Integrated Computerized Deployment System (ICODES) is a fully integrated information system that provides multi-modal load planning capabilities to DOD agencies and services. The combined functionality of ship, air, truck, rail, and yard planning services provides commanders, planners, and operators with a single platform capable of producing and evaluating load plans and alternative actions for various sized units, employing various modes of transportation, in support of peacetime or wartime operations. The ICODES consumes cargo and passenger information from a variety of DOD manifesting systems and, in return, provides load planning, report generation, and forecasting services to USTRANSCOM and its component commands, DOD customers, and other authorized users.

(5) Theater Medical Information Program. The theater medical information program provides a global capability that links medical information data bases to integration centers. These integration centers are accessible to Navy medical personnel and operate in support of Marine forces. The goal is to provide medical integrated automated information using the GCCS and the GCSS, which links all echelons of medical care in support of Marine forces.

(6) Navy Tactical Command Support System. Naval Tactical Command Support System is a multi-application information system program that provides standard information resource management to afloat and shore-based fleet activities. This system was established by the merger of three key programs: the shipboard nontactical automated data processing program, the Naval Aviation Logistics Command Management Information System, and maintenance resource management system (commonly referred to as SNAP, NALCOMIS, and MRMS). The Naval Tactical Command Support System provides a full range of standardized mission support automated data processing hardware and software to support management of logistic

information, personnel, material, equipment maintenance, and finances required to maintain and operate ships, submarines, and aircraft in support of the Navy and Marine Corps.

(7) Support Equipment Resources Management Information System. The support equipment resources management information system (SERMIS) is the primary management information system supporting the aircraft maintenance material readiness list program. Directed by the Office of the Chief of Naval Operation N78, the SERMIS is the single source for baseline budgeting and acquisition of aviation support equipment for Naval Air Systems Command program managers and Marine Corps support equipment logistic managers. The SERMIS provides a centralized and integrated database containing support equipment data for inventory, allowance, rework capability, and production status in a form suitable for online interactive access see MCWP 3-21.2.

(8) Relational Supply. Relational supply (RSupply) provides Navy and Marine Corps personnel the tools and functions necessary to perform their day-to-day business: ordering, receiving, and issuing of services and materials; maintaining financial records; and reconciling supply, inventory, and financial records with the shore infrastructure. The major functions of RSupply are divided into the following subsystems:

- **Site.** Contains information on your own site, serial numbers, user access, validation tables, fund codes, default values, and maintenance data.
- **Inventory.** Provides automated procedures to ensure that physical stock and stock records agree, allowance lists are accurate, usage data is evaluated correctly, and material requirements are anticipated. In addition, it provides programs the ability to balance material requests against available funds and purge storerooms of stock no longer applicable to supported units.
- **Logistics.** Provides automated procedures to create military standard requisitioning and issue procedure (commonly referred to as MILSTRIP) requisitions, receive and store material, issue material to supported and nonsupported customers, process incoming and outgoing supply status, process carcass tracking inquiries and replies, and update all logistic data files.
- **Financial.** Provides automated procedures for assimilating and reporting financial credits and expenditures. Provides an automated reconciliation tool for processing of summary filled order expenditure difference listings manually or through a Standard Accounting, Budgeting, and Reporting System Management Analysis Retrieval System file input as well as aged unfilled order listings.
- **Query.** Provides a real-time automated means of querying data required in decision making, providing status and determining the posture of onboard spares.
- **Interface.** Provides the interfaces required to communicate RSupply information to Organizational Maintenance Management System-Next Generation and Naval Aviation Logistics Command Management Information System (commonly referred to as OMMS-NG and NALCOMIS, respectively) as well as receive data updates.

(9) Total Ammunition Management Information System. The current Marine Corps training ammunition management system is the US Army's Total Ammunition Management Information System (commonly referred to as TAMIS). This system is the automated system

authorized for use by the Marine Corps to forecast and approve munitions requirements, process, and validate requests for both operational and training munitions, and to report expenditure metrics and munitions status. Below are other ammunition management systems—

- **Ordnance Information System-Marine Corps (OIS-MC).** The OIS-MC is an integrated system of application software designed for retail ammunition asset management and reporting. The system is used by all Navy and Marine Corps ashore and afloat activities and contractors holding Navy cognizance ordnance to locally manage ammunition inventory and report to OIS-Wholesale. The OIS-MC can operate as either a stand-alone system or a server network designed to provide automated ammunition requisitioning, status accounting, and management capability at the MALS ammunition supply point level.
- **Marine Corps Ammunition Accounting and Reporting System II (MAARS II).** The MAARS II is the single repository for worldwide status of Marine Corps ground ammunition expendable nonnuclear ordnance requirements, assets, production, expenditures, costs, and technical inventory management data. The MAARS II supports the ammunition management information needs of the stockpile/item managers, the program manager, and Marine forces headquarters. The MAARS II interfaces with other automated information systems (both inter-service and intra-service) to exchange inventory data and related information.

(10) Functional Managers. The MAGTF commander appoints a functional manager for each logistic system. This individual coordinates processing support as well as data collection and distribution with the G-6/S-6. Functional managers for—

- Supply, maintenance, and disbursing systems are managed, controlled, and monitored by the LCE.
- Manpower management systems are managed by the MAGTF manpower staff officer (G-1/S-1).
- Aviation maintenance and flight readiness systems are managed, controlled, and monitored by the ACE.
- Embarkation systems are managed, controlled, and monitored by the MAGTF embarkation officer.

(11) Data Communications. The MAGTF G-6 establishes a data communications network because intra-theater data communications are essential to support high-volume logistic system information exchange requirements. Users not served by the data communications network must use nonelectronic methods to transfer large volumes of logistic data (e.g. external hard drive, compact disc coupled with physical courier). When electronic data communication means are available, nonelectronic backup methods should still be planned.

(12) Information Systems Support Planning. Planning for information systems support must include identification of requirements, establishment of priorities, and allocation of resources. The G-6/S-6 in conjunction with the functional manager must identify the communications and information systems requirements for each major functional system. The information systems management officer then identifies processing priorities and allocates

communication and system resources. The MAGTF OPORD must document the requirements, priorities, and allocations. Also, the OPORD must show the data flow within the MAGTF and between the MAGTF and the defense information systems network data entry point. In addition, the OPORD must depict information systems equipment distribution and maintenance procedures. Ideally, the OPORD references the MAGTF communications and information systems SOP and gives only that supplemental information needed for the specific operation.

3010. LIAISON

Commanders at every level routinely establish contact with other units in their area. At the tactical level, this contact or liaison is established for general operations and logistic support coordination. Liaison improves the LCE's ability to support the supported unit's concept of operations. The LCE staff liaison may include the temporary or permanent assignment of liaison elements to integrate, coordinate, and execute military operations.

a. Liaison Element

The liaison element is the commander's personal representative to another command. These designated liaison elements improve the contact and communications essential to effective command.

(1) Liaison Officer. A liaison officer is the most commonly employed technique for establishing and maintaining close, continuous contact between commands. Use of a single individual with the proper rank and experience conserves manpower while guaranteeing contact.

(2) Liaison Team. A liaison team is assigned to the supported organization when the workload or the requirement for better coordination dictates. Liaison teams normally include a liaison officer, a liaison chief, clerical personnel and/or drivers, and communications personnel with their equipment.

(3) Courier. A courier is "a messenger responsible for the secure physical transmission and delivery of documents and material" *(JP 1-02)*. The courier can function as a liaison element to another command. An experienced, mature courier can amplify information about the situation or issues of concern.

b. Liaison Element Selection Considerations

Although there are no firm rules for selecting liaison personnel, the commander should consider requirements of the task and the individual's—

- Logistic expertise.
- Rank.
- Experience.
- Knowledge.
- Personal initiative.
- Judgment.
- Communications skills.

For LCE units, the requirement for liaison is part of the assigned mission. The formal mission does not specify the type of liaison element to assign in each case. Command liaison should be conducted in all but the most unusual circumstances. The following considerations provide some insights into determining the best type of liaison element to use.

(1) Available Personnel. The lack of qualified personnel may prevent assignment of dedicated liaison elements even where there is a recognized need. If a liaison officer or team is not available, the commander can use couriers. The commander should select only those who have demonstrated the necessary maturity to handle the duties. The overriding consideration is always responsiveness to the supported unit.

(2) Workload. Workload is a variable that influences the commander's decision to provide liaison, as well as the specific type of liaison element. It is a function of the LCE unit's scope of operations, personnel situation, priorities, and time. The workload varies with the size and mission of both the supporting and supported units and can change during the course of an operation. In some situations, the workload may require little more than routine liaison between principal staff officers or their assistants.

(3) Proximity. When units are in proximity, the commander may rely on principal staff officers to maintain effective communications. Conversely, the workload may dictate the use of a dedicated liaison element despite the unit's location.

(4) Tactical Situation. The need for liaison increases as the pace of tactical operations increases. In a static situation, requirements and procedures are routine. As the tempo of operations increases, maintaining liaison becomes more difficult, as well as more critical. Liaison is especially critical during offensive operations and periods of turbulence.

(5) Timeliness. To complement and enhance the desired effects of early logistic planning, liaison elements should be assigned at the first opportunity. Early coordination between combat and LCE units ensures the timely involvement of the LCE units in the planning process.

c. Exchange of Liaison Element

Traditionally, commanders establish liaison from senior to subordinate, supporting to supported, reinforcing to reinforced, and left to right. As with all rules, however, there are situations that dictate exceptions. For example, often situations dictate the exchange of liaison elements between units.

(1) Senior to Subordinate. The assignment of liaison elements within the same command is unusual. The senior HQ would initiate such assignments. As such, HQ must provide the liaison element, with associated support equipment, to the subordinate unit.

(2) Supporting to Supported. The inherent nature of the supporting role normally dictates that the supporting unit provides the liaison element to the supported unit. For task-organized CSS units, the availability of liaison elements depends on the identification of potential liaison requirements during the planning phase. Based on those requirements, the parent command should task-organize the LCE unit with the personnel and equipment to affect liaison.

(3) Reinforcing to Reinforced. Similar-type units reinforce one another. The LCE unit assigned a reinforcing mission provides a liaison element to the reinforced LCE unit.

(4) Left to Right. Traditionally, units on the left flank are responsible for establishing liaison with units on their right. However, LCE units generally do not provide liaison elements to adjacent units. Liaison between the respective commanders and principal staff officers is the norm in such cases.

d. Liaison Element Duties and Responsibilities

Liaison duties and responsibilities closely correlate with those of the G-4/S-4 of the supported unit. The duties are separated into three broad categories:

(1) Advise and/or Assist. The liaison element advises both the supporting commander and the supported commander. It assists the supported unit to determine its requirements, to ascertain associated priorities, and to assign appropriate allocations. The liaison element advises the supported unit on the capabilities of the supporting unit. It assists the supported unit G-4/S-4 to identify those COAs that are most and least supportable from the LCE viewpoint.

(2) Monitor. The liaison element observes the operations of the supported unit and monitors the status of those functional areas in which the parent LCE organization has a concern. Simultaneously, it keeps abreast of the status of its parent organization's operations. Specifically, the element follows activities that affect the capability to provide continuous support.

(3) Coordinate. The liaison element coordinates and expedites the flow of support and information between the two organizations. In this regard, the liaison element serves as the conduit for two-way communications. It is not a substitute for direct coordination between commanders and principal staff officers; rather, it complements and augments such coordination.

e. Liaison Procedures

Initially, the commander of the supporting unit should accompany the selected liaison representatives. This allows the commander to introduce the selected liaison element to the supported commander and staff. This gesture can have a significant long-term impact on the success of subsequent actions with the supported unit. To effectively conduct liaison duties, the element must—

- Become familiar with the capabilities, limitations, and concept of operations of its parent organization before assuming its duties.
- Report to its assigned unit fully prepared to carry out its duties and responsibilities.
- Become familiar with the structure and functions of the supported unit.
- Know the supported unit's mission, concept of operations, and scheme of maneuver.

CHAPTER 4
PLANNING

This chapter describes the planning process and planning products for tactical logistics. In addition, it identifies key factors in each tactical logistic functional area for consideration to help ensure thorough and effective planning. Planning for tactical logistics is concurrent with the larger planning process that prepares the MAGTF for operations.

4001. LOGISTIC PLANNING CONCEPTS

The following basic concepts govern the planning of tactical logistics:

- Logistical planning should be concurrent with operations planning.
- Combat and combat support units should exploit their organic logistical capabilities before requesting assistance from CSS sources.
- The impetus of logistics is from the rear, directly to the using unit.
- The logistic system must be responsive, effective, and efficient.

4002. Planning for Expeditionary Operations

Logistic self-sufficiency is a primary consideration when planning expeditionary operations because MAGTFs are organized to conduct operations in austere environments. Marine forces and MAGTF commanders provide the operational logistic capabilities necessary for conducting expeditionary operations, while tactical logistics are provided by MAGTF commanders and their subordinates. This expeditionary or temporary operations support will be withdrawn after the mission is accomplished. These missions may include—

- Providing foreign humanitarian assistance.
- Providing noncombatant evacuation.
- Conducting peacekeeping operations.
- Countering an act of aggression.
- Countering drug operations.
- Protecting US citizens.
- Defeating an enemy in combat.
- Providing security cooperation.

a. Phases of Action
Expeditionary operations involve five broad phases of action which have strategic, operational, and tactical considerations. See MCDP 3, *Expeditionary Operations*, for additional information.

(1) Deployment. Deployment is the movement of forces to the area of operations (AO). Deployment is initially a function of strategic mobility. Operational-level movement in theater completes deployment as forces are concentrated for tactical employment. Deployment support permits the MAGTF commanders to marshal, stage, embark, and deploy their commands.

Although deployment is a strategic and operational-level concern, tactical-level CSS units (e.g., MLG) may be required to assist the deployment.

(2) Entry. Entry is the introduction of forces onto foreign soil. Normally, entry is accomplished by sea or air, although in some cases forces may be introduced by ground movement from an expeditionary base in an adjacent country. Logistical capabilities are used in the entry phase to develop entry points (e.g., an airfield or port, an assailable coastline, a drop zone, an accessible frontier).

(3) Enabling Actions. These actions are preparatory actions taken by the expeditionary force to facilitate the eventual accomplishment of the mission. Enabling actions may include seizing a port, airfield, or other lodgment for the introduction of follow-on forces and the establishment of necessary logistic and support capabilities. In case of disaster or disruption, enabling actions may involve the initial restoration of order and stability. In open conflict, enabling actions may involve delaying an enemy advance, attacking certain enemy capabilities, or capturing key terrain that is necessary for the conduct of decisive actions.

(4) Decisive Actions. These actions are intended to create the conditions that will accomplish the mission. In disasters, decisive actions may include relief operations. In disruptions, they often include peacemaking and peacekeeping until local government control can be reestablished. In conflict, they usually involve military defeat of the adversary. Logistic organizations provide support across the spectrum of decisive actions.

(5) Departure or Transition. Because expeditions are by definition temporary, all expeditionary operations involve a departure of the expeditionary force or a transition to a permanent presence of some sort. Departure is not as simple as the tactical withdrawal of the expeditionary forces from the scene. It requires withdrawing the force in a way that maintains the desired situation while preserving the combat capabilities of the force. For example, care must be taken to reload the ships of an MPF or MEU to restore their sustainment capabilities because either force may be instantly ordered to undertake another expeditionary operation.

b. Forward-Deployed Logistic Capabilities

The Marine Corps maintains a war reserve program that allows MAGTFs to sustain themselves for a significant period of time during combat operations. Sustainment gives the MAGTFs the required endurance until theater-level supply is established. Sustainment resources that are forward deployed with MAGTFs are augmented and replenished with materiel managed in the war reserve, MPF, and land prepositioning programs. The resulting logistic self-sufficiency is a fundamental, defining characteristic of expeditionary MAGTFs.

(1) War Reserve Materiel. Typically, a combination of nondeployed force-held assets, MPF Class IX offset, and war reserve system programmed purchases will collectively serve to ensure that MEB-level or above MAGTFs can deploy with sufficient equipment and supplies to support up to 60 days of contingency operations. The 60-day level provides reasonable assurance that the force can be self-sustaining until resupply channels are established. Usually, the MAGTF ACE deploys with sufficient aviation-peculiar equipment and supplies for 90 days of contingency operations. Normally, Class V (A) ammunition is not computed in the ACE 90-day sustainment figure due to the large lift requirement associated with Class V (A).

(2) Maritime Prepositioning Force. The MPF is the combination of prepositioned materiel and airlifted elements with a sustainment capability of 30 days. Smaller MAGTFs may be sustained ashore for more or less time depending on the size of the force, the number of MPS in support of that force, and other variables such as inclusion of an aviation logistics support ship (T-AVB).

(3) Geo-Prepositioned Programs. The Marine Corps Prepositioning Program Norway (MCPP-N) is the Marine Corps' land-based geo-prepositioned program. Agreements between the governments of the United States and Norway established the prepositioned MCPP-N stocks, which are used for regional contingencies. The vehicles, equipment, and supplies within the MCPP-N have been configured to support a MAGTF. The stocking goals for a geo-prepositioned program are the same as the MPF ships, although global requirements can be filled with this equipment if directed by HQMC.

c. Marine Expeditionary Planning Organization

The plans and future operations sections prepare plans using the Marine Corps Planning Process (MCPP). See MCWP 5-1, *Marine Corps Planning Process*, for more detail. Future and current operations sections oversee the execution of those plans. Subordinate elements and smaller MAGTFs conduct the same planning; however, their greater focus on the current battle and smaller size may dictate modifications to the staff organization.

(1) Plans Section. Under the staff cognizance of the G-5, the plans section—

- Provides a link between higher headquarters planning sections and future operations section.
- Focuses on deliberate planning and follow-on phases of a campaign or operation.
- Develops branch plans and sequels.

(2) Future Operations Section. Under the cognizance of the G-3/S-3, the future operations section—

- Coordinates with the plans section and current operations sections to ensure integration of the next battle plan.
- Interacts with intelligence collection and the targeting process to shape the next battle.
- Manages the command's planning, decision, execution and assessment cycle to match HHQ battle rhythms and to create the conditions for the success of current operations.

(3) Operational Planning Team. An operational planning team (OPT) is a temporary organization formed around the plans or future operations section to conduct integrated planning. While the current operations section manages the execution of current operations, an OPT plans future operations and develops the operation plan (OPLAN), OPORD, or fragmentary order. The OPT integrates the various staff sections, battlefield function representatives, and subordinate liaisons into the planning process.

(4) Current Operations Section. This section receives the OPORD from future operations and executes the OPORD from the COC. Under the cognizance of the G-3/S-3, the current operations section—

- Coordinates and executes the current order.
- Monitors operations of the MAGTF.
- Prepares fragmentary orders to modify the current OPORD.
- Assesses shaping actions and the progress toward the commander's decisive actions.
- Coordinates terrain management.
- Maintains essential maps and information.
- Provides plans and future operations with situational awareness.
- Provides transition officers to future operations.

4003. TYPES OF JOINT PLANNING

All MAGTF planners must be familiar with JOPES because the Marine Corps continues to operate in a joint or combined environment. Thus, JOPES is how the DOD plans for and conducts joint military operations. As described in JOPES, there are two primary methods of planning joint or combined operations: contingency and crisis action planning. The distinction between the following methods is important because it reflects significant differences in the amount of time available for MAGTF planning:

- Contingency planning (also known as deliberate planning) is conducted principally in peacetime and is accomplished in prescribed cycles that complement other DOD planning cycles. The process requires a significantly longer period of time for completion than crisis action planning.
- Crisis action planning is time-sensitive planning that involves emergencies with possible national security implications.

4004. MARINE CORPS PLANNING PROCESS

The MCPP is the process operating force commanders and their staffs use to provide input to the joint planning process and to plan force organization and employment. Applicable across the range of military operations, the MCPP is designed for use at any echelon of command. It complements joint contingency and crisis action planning procedures outlined in JOPES and provides Marine commanders with a tool for preparing plans and orders. Logisticians participate in all steps of the MCPP with the representatives of the other warfighting functions, staff sections, subject matter experts, and command representatives.

The MCPP establishes procedures for analyzing a mission, developing and analyzing COAs against the threat, comparing friendly COAs against the commander's criteria and each other, selecting a COA, and preparing an OPORD for execution. It organizes the planning process into six manageable, logical steps.

The MCPP provides commanders and their staffs with a means to organize their planning activities and transmit the plan to subordinates and subordinate commands. Through this process, all levels of command begin their planning effort with a common understanding of the mission and commander's guidance. Interactions among various planning steps allow a concurrent, coordinated effort that maintains flexibility, makes efficient use of time available, and facilitates continuous information sharing. See figure 4-1.

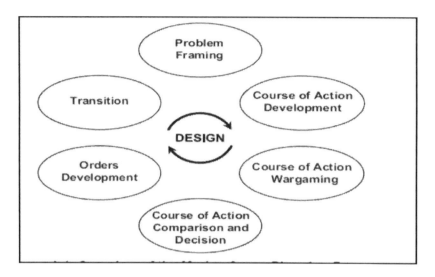

Figure 4-1. The Marine Corps Planning Process.

a. Marine Corps Planning Process and the Logistician

Logistic or support planners may perceive the MCPP as not being logistic-centric or as only used by MAGTF-level staffs. Marine Corps Warfighting Publication 5-1 is written at the MAGTF level, and most formal learning centers teach MCPP with maneuver-centered scenarios, which only require logistic estimates to support maneuver's COAs. Logistics exists to support maneuver, but only learning MCPP through a maneuver lens can result in logisticians with a passive focus on logistic estimates and support procedures. The MCPP is a universally effective analytical decision process that provides a better understanding of capabilities compared to those we are supporting and within the environment we will be operating in. The MCPP is intended to be used as a tool that is scalable to available time and when approaching a complex problem. An analytical decision process is a starting point which helps to inform future intuitive decisions that must be made within extreme time constraints. Therefore, MCPP is malleable to the uniqueness of the organization using it and the mission. Doctrine is not prescriptive because planning products cannot be predictive. The MCPP is the Marine Corps' approach to providing collective understanding and enabling leaders to be proactive during execution. The following paragraphs are intended to provide logistic planners additional tools to apply to their situation and shape their planning efforts.

b. Problem Framing

Problem framing is understood to be the most important step of MCPP because no amount of planning can solve a problem that is insufficiently understood. However, it is important for planners to recognize they will never completely understand a complex problem and the planning process itself will continue to reveal new aspects of a problem. Planners should go into problem framing with the goal of understanding the purpose of their mission, the tasks they have to accomplish towards this purpose and a better understanding of the environment they will operate in.

Problem framing should not be a focus on estimates, but estimates are a critical aspect of problem framing. Currently there are no consolidated logistic estimate tools within the Marine Corps. Logistic planners should look to the Marine Corps Logistics Operations Group or Army Knowledge Online SharePoint sites to identify automated estimate tools. Overlooked or misunderstood elements introduced during problem framing are design and center of gravity (COG) analysis. The concept of design was included when problem framing replaced mission analysis as the first step of MCPP. Design can help prime planners to look up and out at their relationship to a dynamic environment rather than down and in at a perceived static mission. While it is introduced within problem framing, it is in reality a distinct and separate activity that helps inform and guide the planning process and execution of the plan. Outside input to the planning process ensures the decision making process is an open system and accepts the concept of entropy. The most basic explanation of design would be unstructured and honest discussions about problems a unit has or may have and how they can be overcome. Design is a name given to a very natural human phenomenon of sharing information or perspectives in order to develop understanding of the current state and how participants can get to a desired end state. For an example of design as it relates to problem framing, see MCWP 5-1.

Center of gravity analysis is difficult for logistic planners and requires creativity. Often the COG is dismissed as a GCE concept or is consigned to the GCE in an understanding that support activities are often the friendly critical vulnerability. However LCE planners can conduct their own COG analysis to gain a better understanding of how to better apply capabilities to support. A technique is for support planners to understand the enemy does not have to be the conventional enemy maneuver planners often use. A logistic unit may see the enemy as a friendly tank company that is trying to out consume what a direct support CLB can provide. This type of creative thinking enables a creative approach to how planners and commanders use their capabilities against the greatest threat to accomplishing their mission.

c. Course of Action Development

The number of COAs an LCE OPT develops can appear to be challenging. The commander's COA guidance should specify the number of COAs to be developed by the LCE OPT. If multiple COAs are directed, the COAs can be made distinguishable by means of distribution, task organization, or determining if the mission is beyond the functions of logistic support (such as humanitarian assistance/disaster relief or noncombatant evacuation operations). A solid understanding of the physical network analysis (PNA) and how it connects the LCE's support to the MAGTF's end state can greatly assist LCE OPTs in developing distinct and realistic distribution methods or command relationships within task organization.

Discussing COA development requires discussion of the LCE assessment plan. Assessment is a critical activity intended to inform situational understanding during execution, but it begins during COA development. There are three components of assessment:

- Forming a basis for comparison in the form of planning goals (tasks, purpose, conditions, effects, actions).

- Examining the feedback loop that allows us to approximate the situation as it exists and gain an understanding of changes in our environment. This is accomplished by devising collections methods and assessment mechanisms that allow analysis of information.
- Determining the process: Methods to help the commander, via his staff, determine the difference between the operational goals and current reality, the reasons for the difference and ability to recommend change.

The OPLAN/OPORD provides the goals and becomes the basis of comparison. Reporting capabilities and ISR assets provide the reality of the current situation. The devised process provides the feedback loop and assists the commander in understanding where the delta may exist between the desired goals and the reality of the current situation. In addition, the process needs to be able to provide the commander with recommendations. Ultimately, the commander compares whether his planned activities have achieved their desired effects and whether to modify or cease those activities to achieve his desired end state. Assessment plans developed accurately, provide a running comparison of the actual situation in the operational area during execution to the forecasted conditions described in the concept of operations, mission, and commander's intent. Assessment is a continuous activity that encompasses three discreet tasks:

- Gauging strengths, weaknesses, and vulnerabilities.
- Monitoring the situation while measuring the progress of the operation against the desired end state.
- Evaluating the progress of the operation against collectible, observable, relevant (tied to command goals), or measurable metrics.

See MCRP 5-1C, *Operation Assessment*, for more clarification on assessments.

d. Course of Action Wargaming
Wargaming offers challenges to logistic planners because the LCE may not use the traditional enemy to wargame against. Logistics, as a warfighting function, typically does not maneuver against the enemy. Therefore, creativity and critical thinking are required to stress developed COAs. Challenges within the environment, degraded road networks, interrupted supply nodes, unplanned GCE movements, etc. can provide commanders more realistic perspectives of their COAs than physical enemy activity.

e. Course of Action Comparison and Decision
The commander, with the assistance of his principal staff, collects and considers all required information to make an informed decision based on COA comparison.

f. Orders Development
A written order is essential to clearly communicate critical information and provide a common understanding of the unit's problem and goals. The staff should not only focus on "proper" formatting, although important, but on putting critical thought into what must be articulated to properly execute the order. Higher headquarters information must not be regurgitated without purpose and key points developed during the planning process must be written concisely and clearly to be relevant for operations. A common problem for the LCE is that the MCWP 5-1, or any other reference, does not direct how LCEs write annex C or annex D; this is dependent on

commander's preference and the situation. Often in an LCE, annex C is written as logistic operations and annex D as internal logistics, but there are other methods that can be used. The key point is to ensure a discussion takes place between the commander and his staff about how operational logistics are written within the order and the decision is communicated to HHQ, supported, and adjacent units.

The ultimate test of an order's quality is when an officer or staff noncommissioned officer, not exposed to the products planning processes, understands the final product and can apply it.

g. Transition
Transition is the final step of MCPP and may involve a wide range of briefs, drills and rehearsals necessary to ensure a successful shift from planning to executions. Transition is a continuous process that requires a free flow of information between commanders and staff, particularly logistic planners.

4005. CONCEPT OF LOGISTIC SUPPORT

The concept of logistics and CSS is a broad statement of the essential tasks involved in supporting the conduct of MAGTF operations. It gives an overall picture of CSS operations and addresses solutions to shortfalls cited in the CSS estimate. In addition, it is the foundation for subsequent development of detailed logistic and CSS plans and orders by the MAGTF elements. The MAGTF commander's concept for logistics is contained in the MAGTF OPORD, and annex D. The concept for logistics provides guidance for subordinate MAGTF elements and information required for coordination with logistic support agencies external to the MAGTF. The MAGTF G-4/S-4 prepares annex D with input provided by his/her internal staff and input from the LCE G-3/S-3. Subordinate G-4/S-4s conduct detailed planning to accomplish the logistic and CSS tasks promulgated in the OPORD.

4006. PLANNING ELEMENTS

The elements discussed in the following bullets must be addressed in each phase and stage of logistic planning:

- **Mission**. The MAGTF mission is paramount. The missions of subordinate elements must complement the MAGTF mission and may dictate additional parameters for tactical logistic planning.
- **Concept of Operations**. Logistic personnel should fully understand the supported commander's concept of operations. This is vital if they are to anticipate the requirements of the supported organizations. Anticipation is key to the principles of responsiveness and flexibility.
- **Forces**. Available forces and OPLANs dictate logistical requirements. The availability of support from other Services or host nations influences the concept of logistics and CSS. Similarly, enemy capabilities influence the selection of a concept of logistics and CSS in a given situation.
- **Theater Characteristics**. Theater characteristics include the distance between the objective area and sources of supply. Also important is the turnaround time for airlift and

sealift assets. Local populations and environmental conditions (e.g., facilities, road nets, weather, terrain) also affect support operations.
- **Intensity of Operations**. The expected intensity of operations is a key planning factor for quantifying logistic and CSS requirements.
- **Timing and Duration**. The anticipated timing and duration of operations influence planning and preparation, as the time available to complete plans or to procure and stage equipment and supplies may be limited.

4007. PLANNING TECHNIQUES

Limited information and limited time are characteristic of MAGTF planning. Upon receipt of the mission, the MAGTF staff reviews existing OPLANs, SOPs, and joint and Marine Corps lessons learned for related information. Staff members compare plans and SOPs to the assigned mission and to available information at each stage of the planning process. Operational planning often begins with a nucleus staff. During the initial phase, the MAGTF should place particular emphasis on the following techniques:

- **Flexible Approach.** Planning is a continuous process that requires a flexible approach. Initial estimates are based on assumptions and minimal data. Commanders and staffs must continually evaluate previous decisions and guidance. New information can confirm or invalidate previous assumptions or data.
- **Timely Effort.** Logistical planning must begin as early as possible at all levels of command. Early identification of requirements, capabilities, and special considerations accelerates coordination, timely guidance, and essential decisions. As the concept of operations becomes more specific, subordinate elements can begin preparation of more detailed logistic plans.
- **Coordinated Planning.** To accomplish the MAGTF mission, every aspect of the operational concept requires coordination among the GCE, ACE, and LCE. To achieve this, every element has certain responsibilities for logistic planning. This mutual dependence requires concurrent, parallel, and detailed staff planning between and among all elements. Simultaneously, the MAGTF headquarters must coordinate with higher, adjacent, and supporting commands and, possibly, with participating joint and combined staffs. This coordination is essential for integrating MAGTF logistic and CSS operations with those of other organizations.
- **Concurrent and Parallel Development.** Based on both initial and revised guidance, the MAGTF and its elements develop their plans in a concurrent and parallel manner. Integrated planning shortens the planning cycle, enables early identification of potential problems, and improves anticipation of requirements. With proper coordination, concurrent efforts can prevent difficulties that might occur if planning is sequential or isolated. Logistical planning must parallel operational planning. Likewise, the MAGTF concept of operations cannot be developed without full consideration of the supporting concept of logistics and CSS.

4008. DEPLOYMENT PLANNING CONSIDERATIONS

There are two tactical logistic support scenarios to consider when planning for deployment. Under either of the following options, the tactical logistic planner must consider MAGTF requirements in all six CSS functions and their subfunctions:

- The MAGTF can deploy to an area with an established logistic support base. This can be HNS, inter-Service support, or a combination of the two. The logistic planner must plan for reliance on, or expansion of, the existing support base. In addition, the planner must consider an effective alternative to that support if it stops.
- The MAGTF can deploy to an area without an established logistic support base. In this scenario, the logistic planner must rely on organic logistic resources to support the MAGTF.

4009. COMMANDER'S INTENT

Planners cannot foresee every eventuality, and even if they could, plans cannot practically address every possible situation. Commander's intent is the commander's personal expression of the purpose of the operation. Commander's intent helps subordinates understand the larger context of their actions and guides them in the absence of orders. It allows subordinates to exercise judgment and initiative, in a way that is consistent with the higher commander's aims, when the unforeseen occurs. Regardless of the form that it takes, the commander's intent must—

- Be clear, concise, and easy to understand.
- Support the higher, supported commander's intent.
- Include how the commander envisions achieving a decision.
- Provide an end state or conditions that, when satisfied, accomplish the purpose.

4010. OPERATIONAL PLANNING CONSIDERATIONS

Logistical planning focuses on satisfying the logistic requirements generated by the supported force. This planning addresses the estimation of materiel and functional support requirements as well as the organization and employment of organic and supporting tactical logistic organizations. Materiel and functional support requirements are calculated based on experience, assigned missions and tasks, and operational factors (e.g., time available, weather, and enemy).

MAGTF commanders and staff officers should consider the following examples when planning. These examples provide insights for developing and maintaining throughput systems and sustainment capabilities for the execution of logistic support of MAGTF tactical operations.

a. Supply

(1) Ground. Commanders should optimize the basic load for all supplies, including class IX repair parts. The unit's basic load should not exceed the commander's anticipated requirements, even if the unit can carry additional quantities.

(2) Aviation. The squadron maintenance staffs should ensure that their pre-expended bins have been replenished by the supporting MALS. Aviation staffs must coordinate with the supporting MALS, MWSS, and MAG headquarters for aviation-peculiar logistic support en route and within the theater.

b. Maintenance
Each MAGTF element should make maximum use of organic maintenance contact teams (MCTs) and LCE MSTs. Repair and return of equipment as far forward as possible speeds return of equipment to the user. This practice also reduces the burden on both transportation and control capabilities.

c. Transportation
Because transportation is the most limited and limiting logistical capability in the MAGTF, it requires close management. Improper management of transportation assets may degrade combat operations. Supplies should be moved only as needed.

d. External Support
MAGTF plans should make maximum use of HNS and inter-Service support available within the theater of operations. Plans should include, but not be limited to, use of facilities, supplies, utilities, captured materiel, and civilian labor. The LCE commander should keep the number of CSS installations to a minimum and ensure dispersion of installations and capabilities.

e. Forward Support
The farther forward the CSS unit, the less responsibility it should have for routine support tasks. The LCEs should be responsible only for those supplies and services that are critical to combat operations.

f. Air Support
In planning for sustained operations, the MAGTF should expect to receive critical items primarily by air; however, this does not preclude thorough planning for surface lift.

g. Alternate Supply Routes
Transportation planning at every echelon should include the development of alternate supply routes. Use of a single supply route increases the chances that enemy action could severely disrupt or prevent movement.

h. Security
The LCE commander is inherently responsible for the organization's security. While continuing to provide support, the LCE commander must employ both active and passive measures to defend against attempts to disrupt support operations.

4011. Functional Area Planning Considerations

a. Supply
Compromises that are acceptable in peacetime to improve economy and enhance accountability may not be appropriate in a combat situation. For example, storage of a

commodity in a single dump site may be appropriate in peacetime. Centralization in wartime may be unresponsive and reduce survivability. Therefore, the LCE commander may establish multiple CSSAs. Their capacities and locations vary based on the tactical situation, the concept of operations, and the scheme of maneuver.

(1) Supply Cycle. The supply process is a cycle that involves requisition authority, use, and replenishment of supply items. The cycle period for each supply item varies based on criticality code, usage rate, storage and transport capacity, and procurement lead time. Normally, the shorter the cycle, the more intensive the management and transportation effort becomes. Conversely, items with longer cycles require forward planning and more storage area to accommodate the expanded size of the stockage objective.

(2) Phases of Supply Support. The LCE and ACE perform the tactical supply that affects the sustainability of the MAGTF. Tactical supply extends from receipt of finished supplies through issue for use or consumption by the user. The LCE and ACE control the supply process through forecasting, requisitioning, receiving, storing, stock controlling, shipping, disposition, identifying, and accounting procedures established in directives. Ideally, the procedures used in peacetime are the same as those used in wartime. Combat requirements often necessitate rapid processing of requests submitted by unusual methods.

b. Maintenance

Ideally, maintenance procedures should be the same in peacetime and combat but peacetime or garrison maintenance procedures and techniques may not work effectively in combat or field conditions. Maintenance support for Marine aviation has been developed under the Marine aviation logistics support program (MALSP). Logisticians must consider the following factors when planning maintenance systems and procedures:

- Maintenance activities must operate in harsh conditions during tactical operations.
- Limited resources may require around-the-clock work schedules.
- Contamination in the battlespace may further complicate and delay repair of equipment.
- Units must minimize the time required to repair combat essential items. To minimize repair time, units should—
  - Perform only mission essential maintenance during combat. Units must recover, evacuate, and repair equipment as far forward as possible. The lowest level maintenance activity with the proper capability should make the repairs. Repairing equipment as far forward as possible reduces transportation requirements and increases equipment availability.
  - Evacuate inoperable equipment only if they cannot repair it forward or if the repairs will take excessive time. The MAGTF must have a well-defined and understood recovery and evacuation process. In combat, recovery and evacuation may be the most difficult maintenance function. However, this function may also be the most important to sustain the MAGTF's combat power.
  - Make critical repair parts available as far forward as practical. Combat may even require positioning critical parts at the using-unit level. Combat may also dictate greater reliance on selective interchange.

c. Transportation

Transportation planning is throughput planning. It involves the determination of throughput requirements: what, where, when, and how personnel and materiel must move to sustain the force. The transportation planning process is the same regardless of mode, distance, or locale. The operational commander provides requirements and establishes priorities based on the concept of operations. The transportation planner sequences movement requirements in the following order:

- Determine the desired arrival time at destination.
- Select mode of transportation.
- Determine load and pickup points, intermediate and transfer points (as required), as well as off-load and drop points.
- Apply time-distance factors.
- Reconcile conflicting requirements for limited transportation assets (including MHE) and support facilities.
- Test movement plan via route recon or advance party for feasibility.

(1) Planning Elements. The following main elements must be considered when planning transportation:

- **Requirements List.** The requirements list identifies what personnel, supplies, and equipment the planner must move. The planner integrates data from all sources, sequencing it by required delivery date and by priority within the required delivery date. It is further sorted by destination and compiled into a single time-phased listing.
- **Lift Mode.** The selected lift mode identifies what transportation means move the personnel or cargo between the point of origin and destination.
- **Routing.** Routing moves from load and pickup points to intermediate and transfer locations to offload and drop points.
- **Timing.** Timely arrival of personnel, supplies, and equipment at the intended destinations is the goal of transportation planning. The key to transportation scheduling is flexibility. Timing of the beginning and end of each leg of a movement increases flexibility. Basic limitations to timeliness include—
  - Required delivery date at the destination.
  - Time when personnel, supplies, and equipment are available for movement from their points of origin.
  - Time and/or distance factors.
  - Throughput capacities of support facilities.
  - Capacity and security of staging bases and supply depots.
  - Special requirements caused by terrain, climate, and environment.

(2) Planning Process. The transportation planner follows the listed steps when planning for transportation:

- **Determining Requirements.** Each requirement for personnel, equipment, or supplies generates a corresponding requirement for transportation. Transportation planners express initial requirements in terms of tonnage and square footage or the number of personnel and

the distance. The planner estimates requirements based on the supplies needed to support the MAGTF and the average distances during each phase of the operation.
- **Determining Resources.** The transportation planner must consider—
  - Type of transportation units available.
  - Characteristics and capabilities of each mode of transportation.
  - Capabilities of available civilian transportation. (The estimate is based on a survey of facilities, inspection of equipment, and agreements negotiated with civilian transportation operators).
  - Availability of indigenous labor or prisoners of war to supplement personnel resources.
  - Capabilities of available host nation transportation, both civilian and military.
- **Balancing Requirements and Resources.** The balancing process determines whether transportation capabilities are adequate to support the operation. It establishes the workload for each transportation mode. This step is the most time consuming portion of the transportation planning process. Planning must include more than just gross quantities of cargo and transportation resources. It must include planning for C2 and for transportation unit support.
- **Determining Critical Points.** On completing the preliminary plan, the planner has enough information to analyze the transportation system. The planner can identify critical points where bottlenecks can delay throughput. The bottlenecks may occur as a result of shortfalls in either personnel, equipment, or facilities. The planner should also identify critical time periods. Development and analysis of alternative schedules, modes, or routes can alleviate bottlenecks and increase flexibility.
- **Coordination.** Complete coordination is mandatory for integrated transportation support. Original guidance is seldom valid throughout the planning process. Constant coordination is necessary if transportation plans are to change as the commander's concepts, requirements, priorities, and allocations change.

d. General Engineering

The MAGTF engineer assigns and integrates construction tasks and priorities for both Marine and NCF engineer components assigned to the MAGTF. The NCF HQ assists the MAGTF engineer in planning and coordinating construction requirements to best use the unique capabilities of the NCF. Continuous liaison is vital during the planning, deployment, and execution phases of MAGTF operations. The following engineer-support planning areas require special consideration:

- **Heavy Equipment.** Most construction equipment is heavy and slow moving. Though able to negotiate rough terrain, it is too slow to keep up with the supported maneuver forces and must be transported by other assets.
- **Transportation.** Engineer units do not have enough transportation assets to move themselves. When moving a large volume of equipment rapidly or over extended distances, augmentation is necessary.
- **Construction Materials.** Many CSS engineering tasks require large amounts of construction materials. The time, manpower, equipment, and fuel required to assemble and use these supplies are often significant. Careful planning will minimize multiple

handling during movement of these items to the construction site. Movement directly from the source of supply to the job site is optimal.
- **Supply, Maintenance, and Ordnance Support**. Supply, maintenance, and ordnance support for engineer operations is extensive. Engineer units have many low-density items of equipment requiring special maintenance to keep them operational. Low-density items range from mine detectors to stationary pumps and generators to mobile construction equipment. Unique ordnance items include explosive line charges and cratering charges.
- **Utilities Support**. Water purification and power-generating equipment require significant MHE and fuel. Space requirements are normally large, and camouflage is difficult. Utilities installations also generate large amounts of heat and noise.
- **Bulk Fuel Support**. Bulk fuel storage/dispensing requires significant MHE. Space requirements are normally large, and camouflage is difficult.
- **EOD Support**. Movement during tactical operations requires the EOD element to carry large amounts of explosives and other Class V items. The EOD element must have the ability to deploy robotics and other tools to ensure mobility and force protection.

e. Health Services

Commanders are responsible for the health and welfare of their troops. When planning, the terms "role" and "level" are used interchangeably. The equipment for some MAGTF medical units requires external transportation, fuel, and utilities support to be fully operational capable. Additional guidance for medical services can be found in the CJCSM 3122.03, *Joint Operation Planning and Execution System.*

Although HSS staffs conduct medical planning within the MAGTF, logistic planners should ensure—

- Complimentary equipment and associated consumables kits (AMAL and ADAL) are in sufficient quantities to support the force.
- Narcotics handling and security procedures are established.
- Medical regulating channels and procedures for the movement and tracking of casualties between and within the levels of care are established.
- The mix of dedicated versus opportune lift for casualty evacuation is decided based on the concept of operations, casualty estimates, and METT-T.
- Role I-III treatment facilities are identified, and medical evacuation policies are established.
- Preventive medicine requirements and preventive medicine technicians for insect control and redeployment agriculture inspections are identified.
- Mass casualty procedures are established.
- Primary and secondary casualty receiving and treatment ships (CRTSs) are identified for amphibious operations.
- Force health protection and special medical treatment requirements for the AO (e.g., immunizations, antivenin, blood products, and anti-malarial medication) are identified.
- Plans for the disposal and management of medical waste are developed.

f. Services

Planning considerations for services vary for each particular service's function and the operational situation. The following factors are common to all services' functions:

- **Responsibility.** Units are responsible for executing command services' functions consistent with the organic capabilities specified in their TO mission statement. Equipping and manning of detachments should be consistent with this specification. Higher-echelon organizations are responsible for augmenting or reinforcing subordinate unit capabilities. The LCE provides CSS services' functions to the MAGTF elements, as directed by the MAGTF commander.
- **Chain of Command.** Combat service support services' functions are typically implemented in operational chains of command. In contrast, most command services' functions normally operate in administrative chains of command in garrison and may continue to do so even after deployment. Element commanders must consider problems that deployments might pose for continuing administrative support when preparing plans for command services' functions. When appropriate, specific guidance should be issued for shifting command services' functions to the operational chain of command and processing these functions via staff cognizance of the MAGTF CE.

4012. COORDINATING SUPPORT

Effective logistical planning requires a coordinated effort between the supported force and the sup- porting organizations. Both supported and supporting organizations make planning and subsequent support operations more efficient through careful calculation of requirements over specified periods of time while coordinating to reconcile potential shortages or excesses. Ground-common and aviation-peculiar logistic support must be provided in the right quantity, at the right time, and in the right place. Providing too much materiel or too robust a service at one location may disrupt operations of the supported unit or deprive other supported units of what they need when they need it. Effective planning can minimize the occurrence of shortages or excesses. Supported organizations must—

- Calculate their requirements as precisely as possible.
- Factor organic or attached and/or direct support cargo and personnel transportation capacity into the requirements calculation.
- Prioritize requirements.
- Integrate requirements with expected schedule and duration of the operation.
- Verify critical materiel or services allocations made by higher authority when determining requirements for tactical missions.

Supporting organizations must—

- Review with the supported organization the support requirements as they are developed.
- Coordinate with the supported organization to refine the requirements based on the supported organization's competing requirements.

- Procure materiel and task-organize internally to provide support efficiently.
- Plan support distribution by anticipating demand.
- Provide the support required.

4013. INTELLIGENCE SUPPORT

Intelligence information is essential for planning tactical logistic operations. Logistic intelligence is specific intelligence information that assists logistic organizations in accomplishing their assigned missions. It focuses on the infrastructure in the area of interest and on how the weather, enemy, and terrain would affect tactical logistic operations. Logistic intelligence is a product of the MAGTF's intelligence cycle and intelligence preparation of the battlespace (IPB) functions. The following IPB products are typically of interest to logisticians and may be categorized as a PNA:

- LOC and route studies.
- Port and harbor studies.
- Airfield studies.
- Drop zone and landing zone studies.
- Bridge and inland waterway studies.
- Key facilities and targets overlays.
- Specialized weather and terrain studies.
- Modified combined obstacle overlay.
- Available resources.
- Natural (lumber, water, construction materials, food).
- Manmade (construction equipment, rail networks, power grids, sewer systems, sanitary dumps, enemy logistic capabilities, etc.).
- Medical threats to the health of the force.

4014. HOST-NATION SUPPORT

When feasible, MAGTF plans should make maximum use of HNS available within the theater of operations. However, HNS is not a substitute for essential MAGTF organic tactical logistic and CSS capabilities. Host-nation support can augment MAGTF capabilities preserving the MAGTF's accompanying supplies, which can increase the length of time that accompanying supplies can support the operation. Bilateral (between the United States and a single country) and multilateral (among members of a coalition, such as the North Atlantic Treaty Organization [NATO]) host-nation support agreements can be an integral part of sustainment planning. Normally, host-nation support agreements are prepared at the operational level through JTF or combatant commander. Implementation of existing agreements and/or preparation of new agreements must be coordinated between the MAGTF CE and the appropriate higher authority in the US chain of command. Plans should include, but not be limited to, use of facilities, supplies, utilities, captured materiel, and civilian labor.

4015. PLANNING DOCUMENTS

The logistic/CSS estimate, annex D of the OPORD (concept of logistics and CSS), and the LCE OPORD are the primary MAGTF tactical logistical planning documents. Table 4-1 summarizes the standard logistic planning documents and identifies the preparer.

Table 4-1. Logistic and CSS Planning Documents.

Document	Prepared By
Logistic/Combat Service Support Estimate	CE, GCE, ACE, LCE down to battalion and squadron level
Annex D (Logistics/Combat Service Support) to OPORD	CE, GCE, ACE, LCE down to battalion and squadron level
LCE Operation Order	LCE

a. Logistic/Combat Service Support Estimate

The estimate is a rapid assessment by the G-4/S-4 of logistic capabilities and limitations for each proposed COA. It analyzes the COAs under consideration to provide the logistical aspects of relative combat power. The estimate helps determine the most desirable and most supportable COA from the CSS standpoint. Additionally, this document provides the basis for later planning. See appendix B of this publication for a sample of the logisticsCSS estimate.

The commander decides which COA will be used to accomplish the assigned mission. As an advisor, the G-4/S-4 provides the commander with information and makes recommendations based on the logistic/CSS estimate.

The logistic/CSS estimate is the result of an examination of the logistical factors which influence contemplated COAs and an appraisal of the degree and manner of that influence. The estimate looks at the six tactical logistic functional areas. The estimate compares requirements, available assets, problems, limitations, advantages, and disadvantages for each COA. The logistic/CSS estimate assesses the limitations of each COA. It also determines what actions are necessary to overcome any problems or limitations. If any COA is not supportable, the estimate specifically states this. It gives the commander enough information to make a decision based on the suitability, feasibility, acceptability, and relative merit of each COA from a logistic standpoint.

b. Annex D to the Marine Air-Ground Task Force Operation Order

Annex D reflects the commander's plans, guidance, and directions for employment of logistic capabilities. This annex complements the concept of operations and amplifies paragraph 4 of the OPORD (Administration and Logistics). Annex D begins with the concept of operations and the supporting concept of logistics. It assigns tasks and responsibilities for logistics and CSS among the elements in each functional area. It also identifies support required from external agencies. Finally, it provides guidance and information (such as priorities and allocations) for planning, coordinating, and executing MAGTF logistic operations. See appendix C of this publication for a sample Annex D (Logistics/Combat Service Support).

(1) Commander's Guidance. Annex D promulgates the commander's overall plan and guidance for the provision of logistic support to the MAGTF during each phase of the operation. This annex specifies those requirements, priorities, and allocations that are necessary for the integration of the logistic effort in support of the MAGTF. It includes deployment, employment, sustainment, and redeployment planning matters. It includes external support coordination requirements and internal employment directives to present a single, unified plan for logistic support.

(2) Concept of Logistics and Combat Service Support. The concept of logistics and CSS (paragraph 3a of annex D) is a broad statement of the essential logistic and CSS tasks involved in supporting the concept of operations. It is the basic unifying foundation for subsequent development of detailed logistic and CSS plans and orders by the MAGTF elements.

(3) Staff Responsibility. The MAGTF G-4/S-4, in coordination with other staff sections, the subordinate G-4/S-4s, and LCE G-3/S-3, prepares annex D. This document also contains the specific requirements, priorities, and allocations for logistics and CSS to support the concept of operations and scheme of maneuver. Each subordinate organization down to the battalion and squadron level publishes an annex D. Optionally, they may use paragraph 4 of the OPORD to provide logistic guidance to subordinate units. The use of references to local SOPs contributes to sound plans and helps to avoid unnecessarily lengthy and detailed OPORDs.

(4) Concept of Aviation Logistic Support. Aviation logistic support is addressed in the aviation estimate of supportability and Appendix 10 (Aviation Logistic Support) to annex D to the OPORD.

c. Logistics Combat Element Operation Order

The LCE OPORD states the mission of the LCE, establishes task organizations, and assigns missions to each subordinate unit. It also states the LCE commander's requirements, priorities, and allocations for accomplishing the mission.

The LCE OPORD amplifies information normally contained in SOPs concerning CSS provided to other MAGTF elements. Primarily, the OPORD provides specific guidance and direction to subordinate CSS units regarding their tasks and missions. The LCE G-3/S-3 is responsible for preparing the LCE OPORD. The LCE G-4/S-4 prepares annex D to the LCE OPORD.

d. Standing Operating Procedures

Standing operating procedures are a set of instructions that can be standardized and that are applicable unless ordered otherwise. An SOP is a general order that deals with tactical and administrative procedures not covered by regulatory or doctrinal publications.

The recurrent nature of logistic functions lends them to procedural standardization. As such, SOPs contribute to simplicity, clarity, and brevity. Reliance on SOPs in the various CSS planning documents simplifies and shortens those documents. It is not necessary to list SOPs as references; however, the order should cite the SOPs in the body of the document.

In addition to their advantages in the preparation of planning documents and orders, SOPs improve support by promoting familiarity and mutual confidence between supported and supporting units and personnel. They also reduce the confusion often associated with combat conditions.

e. Other Planning Documents

The G-4/S-4 has staff cognizance for major input to other documents. Many of these documents are unique to landing force operations. Other doctrinal publications, such as JP 3-02.1, *Amphibious Embarkation and Debarkation,* and MCWP 3-31.5, *Ship-To-Shore Movement,* discuss the following documents in detail:

- Embarkation plan.
- Plan for landing supplies.
- Landing plan.
- Organization for embarkation and assignment to shipping tables.

APPENDIX A
LOGISTIC AND COMBAT SERVICE SUPPORT TASK-ORGANIZATION GUIDE

Function	Capabilities			
	CE	ACE	GCE	LCE
Supply Requirements Procurement Storage Distribution Salvage Disposal Requisition authority	Capable of internal ground supply tasks	Group/squadrons capable of internal ground supply tasks MALS performs aviation supply tasks	Regimental headquarters, battalions, and separate companies capable of internal ground supply tasks	Battalions capable of internal ground supply tasks Supply battalion provides ground supply support for the MAGTF
Maintenance Inspection and classification Service, adjustment, tuning Testing and calibration Repair Modification Rebuilding and overhaul Reclamation Recovery and evacuation	Capable of authorized maintenance tasks on assigned ground equipment	Groups/squadrons capable of authorized field level maintenance for assigned ground equipment Squadrons perform organizational maintenance on assigned aircraft MALS performs intermediate and limited depot maintenance on supported aircraft	Organizations capable of authorized maintenance tasks Perform organizational- and intermediate-level maintenance on assigned ground equipment	Battalions capable of authorized maintenance tasks Perform organizational- and intermediate-level maintenance on assigned ground equipment Maintenance battalion provides field and limited depot-level maintenance support for designated MAGTF ground equipment Medical battalion performs maintenance on MAGTF class VIII (medical) materiel

Function	Capabilities			
	CE	ACE	GCE	LCE
Transportation Embarkation Landing support Port and terminal operations Motor transport Aerial delivery Freight/passenger transportation MHE	Capable of preparing assigned personnel, equipment, and supplies for aerial and/or surface embarkation; limited capability for ground transport using organic light and medium trucks	General capability for preparing assigned personnel, equipment, and supplies for aerial or surface embarkation Capable of managing terminal operations and providing aerial transport for selected passengers and cargo Most organic ground transport centralized in the MWSSs	Capable of preparing assigned personnel, equipment, and supplies for air and/or surface embarkation Limited capability for ground transport using organizational light and medium trucks, reinforced as necessary with medium trucks from the truck company of the H&S battalion	Capable of preparing assigned personnel, equipment, and supplies for aerial and/or surface embarkation Transportation support battalion provides landing support, aerial delivery, port and terminal operations, medium- and heavy-truck transportation of freight and passengers, and MHE
General Engineering Engineer reconnaissance Bridging Bulk fuel Construction Facilities maintenance Demolition/obstacle removal EOD Water production and storage Power generation and distribution	Limited organic capability, focused on establishing and running field command posts	The ACE capability for general engineering tasks is centralized in the MWSSs and focused on support of airfield operations	Limited organic engineering capability for combat support tasks is centralized in the combat engineer battalion	Engineer support battalion provides MAGTF capabilities for general engineering tasks and can also reinforce MWSSs and the combat engineer battalion if necessary
Health Services Casualty management Force health protection and Prevention Medical logistics Medical command and control Medical stability operations	Limited organic capability for health care and unit-level casualty care and evacuation	Organic capability for health care and unit-level casualty care and evacuation in separate squadrons and groups Aviation medical services available in aircraft groups	Organic capability for health maintenance and unit-level casualty care and evacuation in regimental headquarters, battalions, and separate companies	Organic capability for health maintenance and unit-level casualty care and evacuation in all battalions Medical battalion provides shock-trauma and surgical support to the MAGTF Dental battalion provides dental services for the MAGTF

Function	Capabilities			
	CE	ACE	GCE	LCE
Services *CSS* Disbursing Postal services Exchange services Security support Legal services Civil affairs Mortuary affairs Contracting	Limited organic capabilities for postal, security, and legal services	Appropriate organic capabilities for disbursing, postal, security, and legal services; civil affairs; mortuary affairs in separate squadrons and groups	Appropriate organic capabilities for disbursing, postal, security, and legal services, in regimental headquarters and battalions	Appropriate organic capabilities for disbursing, postal, security, and legal services; civil affairs; and mortuary affairs in headquarters elements H&S battalion provides additional support in all services to the MAGTF
Command Personnel administration Religious ministries Financial management Information services Communications Billeting Food Services Band Morale, welfare, and recreation (MWR)	Capable of organic command support functions for assigned personnel and organizations; at the MEF level the CE may be required to organize, train, and equip a band	Capable of organic command support functions for assigned personnel and organizations; at the MAW level the ACE may be required to organize, train, and equip a band	Capable of organic command support functions for assigned personnel and organizations; at the division level the GCE may be required to organize, train, and equip a band	Capable of organic command support functions for assigned personnel and organizations; normally the MLG will not be tasked with organizing, training, and equipping a band

Notes:

1. The CE and the GCE contain organic capabilities for internal ground logistic functions per applicable TOs and TEs.

2. The ACE contains organic capabilities for internal ground-common and aviation-peculiar logistic functions per applicable TOs and TEs.

3. The LCE contains organic capabilities for both internal and external (i.e., MAGTF support) ground logistic functions in accordance with the applicable TOs and TEs.

APPENDIX B
SAMPLE FORMAT OF A LOGISTIC/COMBAT SERVICE SUPPORT ESTIMATE

CLASSIFICATION

Copy no. ____ of ____ copies
OFFICIAL DESIGNATION OF COMMAND
PLACE OF ISSUE
Date/time group
Message reference number

LOGISTIC/COMBAT SERVICE SUPPORT ESTIMATE (U)

(U) REFERENCES: As appropriate to the preparation of the estimate.

1. (U) Mission

    a. (U) Basic Mission. State the mission of the command as a whole.

    b. (U) Purpose of the Estimate

        (1) (U) Determine if combat service support (CSS) capabilities are sufficient to support proposed courses of action (COAs).

        (2) (U) Determine which COA is most desirable from a logistic and/or CSS standpoint.

        (3) (U) Determine what measures must be taken by the commander to overcome logistic and/or CSS problems and/or limiting factors in supporting each COA.

2. (U) Situations and Considerations

    a. (U) Enemy Forces

        (1) (U) Present Disposition of Major Elements. Refer to the Intelligence Estimate.

        (2) (U) Major Capabilities. List enemy capabilities that are likely to affect friendly logistic and/or CSS matters.

Page number
CLASSIFICATION

CLASSIFICATION

    (3) (U) <u>Other Capabilities and/or Limitations</u>. List enemy capabilities and/or weaknesses that are likely to affect logistics and/or CSS or tactical situation.

b. (U) <u>Own Forces</u>

    (1) (U) <u>Present Disposition of Major Combat and Combat Support Elements</u>. May be shown as a situation map or an overlay appended as an annex with reference to the aviation combat element (ACE) logistic/CSS estimate by including the statement "See also Aviation Combat Element Logistic/CSS Estimate."

    (2) (U) <u>COAs</u>. State the tactical COAs that are under consideration.

c. (U) <u>Characteristics of the Area</u>. List those characteristics that are likely to affect the logistic and/or CSS situation such as weather, terrain, hydrography, communications routes, and local resources.

d. (U) <u>Current Logistic and/or CSS Status</u>. Give a brief description of the current logistic and/or CSS status, including any planned or known changes before and during the period covered by the estimate. The following sub-paragraphs address typical CSS areas of concern. If possible, state specific quantities.

    (1) (U) <u>CSS Organizations and Task Organizations</u>. Each organic CSS organization or task organization is described using the following format:

        (a) (U) <u>Locations</u>. May be an overlay.

        (b) (U) <u>Missions and/or Tasks</u>.

        (c) (U) <u>Task Organizations and Command Relationships</u>.

        (d) (U) <u>General Capabilities and Status</u>. Capabilities and status are described in terms of task organization using the applicable categories listed in paragraphs (2) through (13) below.

        (e) (U) <u>Tactical Responsibilities</u>. List if any.

        (f) (U) <u>Communications and Automated Data Processing Systems Support Arrangements</u>.

Page number
CLASSIFICATION

CLASSIFICATION

(2) (U) <u>Personnel</u>

    (a) (U) <u>Strengths</u>. Identify strengths of each major subordinate unit.

    (b) (U) <u>Replacements</u>. Identify replacements on hand, replacements to be received, and the quality of the replacements.

    (c) (U) <u>Morale</u>. Determine the level of fighting spirit, significant factors affecting current morale, religious and welfare matters, and awards.

    (d) (U) <u>Personal Services Support</u>. Identify the required exchange, postal, recreational, and special services support.

    (e) (U) <u>Military Justice</u>. Identify court martial and correction facilities.

    (f) (U) <u>Personnel Procedures</u>. List significant items, if any.

(3) (U) <u>Supply</u>. Identify procurement, storage, distribution, and salvage methods.

(4) (U) <u>Maintenance</u>. Determine management, operations, and workload.

(5) (U) <u>Transportation</u>. Identify motor transport, helicopters, amphibious vehicles, and cargo aircraft; motor transport convoy control; and main supply routes.

(6) (U) <u>Engineer Support</u>. Identify construction and maintenance of roads, bridges, airfields, helicopter landing sites, bulk fuel sites and pipelines, camps, and utilities (including bath, fumigation, laundry, electrical power, and water points).

(7) (U) <u>Landing Support</u>. Identify beach, landing zone, and aerial delivery support operations.

(8) (U) <u>Medical and/or Dental</u>. Identify preventive medicine, casualty collection, evacuation (including evacuation policy), and hospitalization support.

(9) (U) <u>Military Police</u>. Identify the number on hand and evacuation procedures for prisoners of war, the straggler rates and control, and the traffic control methods.

Page number
CLASSIFICATION

CLASSIFICATION

 (10) (U) <u>Civilian Employees</u>. Identify the number, restrictions on use, and organization of civilian employees.

 (11) (U) <u>Civil Affairs Support</u>. Identify CSS of the civil affairs effort.

 (12) (U) <u>Automated Data Processing Systems</u>. Identify management, operations, and support of command automated data processing systems support.

 (13) (U) <u>Miscellaneous</u>. Identify food services, material handling equipment, and financial management (disbursing, budgeting, and accounting) methods.

 e. (U) <u>Assumptions</u>. State those assumptions made for the preparation of this estimate. An example of the critical assumption is the estimation of the length of time for the entire operation and for each COA (if different).

 f. (U) <u>Special Factors</u>. List items not covered elsewhere, such as state of training of CSS personnel or task organizations.

3. (U) <u>Analysis</u>. Under each of the following categories, analyze each COA that is under consideration in light of all significant factors to determine problems that may arise, measures necessary to resolve those problems, and any limiting factors that may exist.

 a. (U) <u>Course of Action #1</u>

  (1) (U) <u>Logistic and/or CSS Organizations and Task Organizations</u>. Describe each organic logistic and/or CSS organization or task organization using the following format.

   (a) (U) <u>Locations</u>. May be an overlay.

   (b) (U) <u>Missions and/or Tasks</u>.

   (c) (U) <u>Task Organizations and Command Relationships</u>.

   (d) (U) <u>General Capabilities and Status</u>. Describe capabilities and status in terms of task organization using the applicable categories listed in paragraphs (2) through (13) below.

Page number
CLASSIFICATION

CLASSIFICATION

 (e) (U) <u>Tactical Responsibilities</u>. If any.

 (f) (U) <u>Communications and Automated Data Processing Systems Support Arrangements</u>.

(2) (U) <u>Personnel</u>

 (a) (U) <u>Strengths</u>. Identify the strengths of each major subordinate unit.

 (b) (U) <u>Casualties</u>. Determine expected casualties for this COA.

 (c) (U) <u>Replacements</u>. Identify replacements on hand, replacements to be received, and the quality of the replacements.

 (d) (U) <u>Morale</u>. Identify significant factors affecting current morale, religious and welfare matters, and awards.

 (e) (U) <u>Personal Services Support</u>. Identify exchange, postal, and recreation and/or special services support.

 (f) (U) <u>Personnel Procedures</u>. List significant items, if any.

(3) (U) <u>Supply</u>. Identify procurement, storage, distribution, and salvage methods.

(4) (U) <u>Maintenance</u>. Identify management, operations, and workload.

(5) (U) <u>Transportation</u>. List motor transport, helicopters, amphibious vehicles, and cargo aircraft; motor transport convoy control; and main supply routes.

(6) (U) <u>Engineer Support</u>. Identify construction and maintenance of roads, bridges, airfields, helicopter landing sites, bulk fuel sites and pipelines, camps, and utilities (including bath, fumigation, laundry, electrical power, and water points).

(7) (U) <u>Landing Support</u>. Identify beach, landing zone, and aerial delivery support operations.

(8) (U) <u>Medical and/or Dental</u>. Identify preventive medicine, casualty collection, evacuation (including evacuation policy), and hospitalization support.

Page number
CLASSIFICATION

CLASSIFICATION

   (9) (U) <u>Military Police</u>. Identify the number on hand and evacuation procedures for prisoners of war, the straggler rates and control, and the traffic control methods.

   (10) (U) <u>Civilian Employees</u>. Identify the number, restrictions on use, and organization of civilian employees.

   (11) (U) <u>Civil Affairs Support</u>. Identify the CSS of the civil affairs effort.

   (12) (U) <u>Automated Data Processing Systems</u>. Identify management, operations, and command automated data processing systems support.

   (13) (U) <u>Miscellaneous</u>. Identify food services, material handling equipment, and financial management (disbursing, budgeting, and accounting) methods.

  b. (U) <u>Course of Action #2</u>. Same subparagraphs as shown for COA #1.

  c. (U) <u>Course of Action #3</u>. Same subparagraphs as shown for COA #1.

4. (U) <u>Evaluation</u>. Based on the foregoing analyses, summarize and compare the advantages and disadvantages of each COA under consideration from a logistic and/or CSS standpoint.

5. (U) <u>Conclusion</u>

  a. (U) <u>Preferred Course of Action</u>. State which COA, if any, can best be supported from a logistical and/or CSS standpoint.

  b. (U) <u>Major Disadvantages of Other Courses of Action</u>. State whether any or all of the remaining COAs can be supported from a logistical and/or CSS standpoint, citing the disadvantages that render them less desirable.

  c. (U) <u>Logistic and/or CSS Problems and Limitations</u>. Cite any significant logistical and/or CSS problems to be resolved and any limitations to be considered in each COA.

  d. (U) <u>Decision or Action</u>. State those measures that are necessary to resolve those logistical and/or CSS problems cited above.

/s/ _____

ANNEXES: (As required)

Page number
CLASSIFICATION

Appendix C
Sample Format of Annex D
(Logistics/Combat Service Support)

CLASSIFICATION

Copy no.___of___copies
OFFICIAL DESIGNATION OF COMMAND
PLACE OF ISSUE
Date-time group
Message reference number

ANNEX D TO OPERATION ORDER OR PLAN (Number) (Operation CODE WORD) (U)
LOGISTICS/COMBAT SERVICE SUPPORT (U)

(U) REFERENCES: Cite references necessary for a complete understanding of this annex.

1. (U) Situation

    a. (U) Adversary. Refer to Annex B (Intelligence). Provide available information on adversary actions or intent to conduct actions to disrupt or degrade envisioned friendly logistic and combat service support operations. Include information on adversary capabilities or assets that can augment friendly logistic and combat service support operations.

    b. (U) Friendly. List supporting logistic or combat service support organizations not subordinate to the force and the specific missions and tasks assigned to each.

    c. (U) Infrastructure. Refer to Annex B (Intelligence). Provide information on existing infrastructure, such as ports, factories, fuel and water sources, and lines of communications that can be used to support friendly logistic and combat service support operations.

    d. (U) Attachments and Detachments. Refer to Annex A (Task Organization). List logistic and combat service support units from other Services/nations attached to the force. List all Marine Corps logistic and combat service support units detached to support other friendly forces.

    e. (U) Assumptions. State realistic assumptions and consider the effect of current operations on logistic capabilities. Omitted in orders.

Page number
CLASSIFICATION

MCTP 3-40B. Tactical-Level Logistics

CLASSIFICATION

 f. (U) <u>Resource Availability</u>. Identify significant competing demands for logistic resources where expected requirements may exceed resources. Include recommended solutions within resource levels available for planning, if any, and reasonably assured HNS.

 g. (U) <u>Planning Factors</u>. Refer to and use approved planning factors and formulas, except when experience or local conditions dictate otherwise. When deviating from planning factors, identify the factors and the reason.

2. (U) <u>Mission</u>. Provide the command's mission from the base order.

3. (U) <u>Execution</u>

 a. (U) <u>Concept of Logistics and Combat Service Support</u>. State the concept for logistics and combat service support operations necessary to implement the order or plan. Describe how the logistic and combat service support assets will be organized and positioned to execute the mission. The concept may include planned employment of other Service and nation logistic and combat service support forces, HNS logistic capabilities, or operation of the lines of communications.

 b. (U) <u>Tasks</u>

  (1) (U) Assign logistic and combat service support responsibilities to subordinate logistic organizations.

  (2) (U) Identify and assign responsibility for logistics and combat service support required from other commands, Services, or nations.

  (3) (U) Identify and assign responsibility for logistics and combat service support required for forces assigned or attached from other commands, Services, or nations.

  (4) (U) Identify and assign responsibility for logistics and combat service support required for Marine Corps forces assigned or attached to other commands, Services, or nations.

  (5) (U) Assign responsibilities to support joint boards and committees, such as transportation and procurement, and other Services or nations providing services.

4. (U) <u>Administration and Logistics</u>

 a. (U) <u>Logistics and Combat Service Support</u>

Page number
CLASSIFICATION

CLASSIFICATION

(1) (U) <u>Supply</u>. Refer to Appendix 7 (Supply). Summarize the following, in coordination with supporting commanders and Service component commanders, if different from standard planning factors. Place detailed discussions in the appendices and listings of supply depots, terminals, and lines of communications in tabs or the appropriate appendices.

    (a) (U) <u>Distribution and Allocation</u>

        <u>1</u> (U) Purpose, location, and projected displacement of main and alternate supply depots or points and supporting terminals and ports to be used or considered.

        <u>2</u> (U) Prepositioned logistic resource allocation.

        <u>3</u> (U) Existing terminals and lines of communications and the known or estimated throughput capability. Indicate the time-phased expansion necessary to support the plan.

    (b) (U) <u>Level of Supply</u>

        <u>1</u> (U) Indicate the time-phased operating and safety levels required to support the plan.

        <u>2</u> (U) Indicate the prepositioned war reserve materiel requirements to support the time-phased deployments pending resupply.

        <u>3</u> (U) Specify significant special arrangements required for materiel support beyond normal supply procedures.

        <u>4</u> (U) Indicate anticipated shortfalls.

        <u>5</u> (U) Indicate common user logistic supply support responsibilities and arrangements.

    (c) (U) <u>Salvage</u>. Provide instructions for and identify the logistic impact of the collection, classification, and disposition of salvage.

    (d) (U) <u>Captured Adversary Materiel</u>. Provide instructions for the collection, classification, and disposition of adversary materiel. See Annex B (Intelligence) for further guidance. See appendix 10 to Annex B (Intelligence) for specific instructions for the disposition of captured adversary cryptographic equipment.

Page number
CLASSIFICATION

CLASSIFICATION

    (e) (U) <u>Operational Contract Support</u>. See MCRP 4-11H, *Multi-Service Tactics, Techniques, and Procedures for Operational Contract Support* .

        <u>1</u> (U) Identify acquisition of goods and services in the following categories:

            <u>a</u> (U) The general categories of materiel and services that are available and contemplated as a supplement to regular sources.

            <u>b</u> (U) Those that may be used as emergency acquisition sources.

        <u>2</u> (U) Make a statement concerning the dependability of the local acquisition or labor source in each of the aforementioned categories and the joint or Service element that will obtain or manage these resources.

        <u>3</u> (U) State that all essential contractor services, to include new and existing contracts, have been reviewed to determine which services will be essential to OPLAN execution. Make a statement concerning the existence of contingency plans to ensure the continuation of these essential services.

    (f) (U) <u>Petroleum, Oils, and Lubricants</u>. Refer to Appendix 1 (Petroleum, Oils, and Lubricants Supply).

(2) (U) <u>External Support</u>. Refer to Appendix 11 (External Support). Provide the required planning information including type and quantity of support and instructions where inter-Service and cross-Service arrangements for common supply and service support are appropriate.

    (a) (U) Summarize major support arrangements that are presently in effect or that will be executed in support of the plan.

    (b) (U) Include significant inter-Service and cross-Service support arrangements. Refer to appropriate annexes or appendices.

    (c) (U) Include foreign and HNS.

(3) (U) <u>Maintenance</u>

    (a) (U) <u>General</u>. Refer to Appendix 12 (Maintenance).

    (b) (U) <u>Specific Guidance</u>

Page number
CLASSIFICATION

CLASSIFICATION

        <u>1</u> (U) Include sufficient detail to determine the requirements for maintenance facilities needed to support the plan.

        <u>2</u> (U) Indicate the level of maintenance to be performed and where it is to occur, including host nation or contractor facilities, if applicable.

(4) (U) <u>Transportation</u>

    (a) (U) <u>General</u>. Refer to Appendix 4 (Mobility and Transportation). Provide general planning or execution guidance to subordinate and supporting organizations to facilitate transportation of the force and its sustainment. This can include movement and use priorities.

    (b) (U) <u>Mobility Support Force and Movement Feasibility Analysis</u>. Provide an estimate of the mobility support and movement feasibility of the plan. Include in the analysis any appropriate remarks affecting mobility and transportation tasks. Consider the availability of adequate lift resources for movements of personnel and equipment, airfield reception capabilities, seaport and aerial port terminal capabilities, and port throughput capabilities. Also, consider any features that will adversely affect movement operations, such as the effect of deployment or employment of forces and materiel on airfield ramp space (to include possible HNS).

(5) (U) <u>General Engineering Support Plan</u>. Refer to Appendix 13 (General Engineering). State the rationale if Appendix 5 (Civil Engineering Support Plan) is not prepared. Indicate the general engineering support activities applicable to the basic operation order or plan and the policies for providing these services.

(6) (U) <u>Health Services</u>. Refer to Appendix 9 (Health Services).

(7) (U) <u>Services</u>. Refer to Appendix 8 (Services).

(8) (U) <u>Mortuary Affairs</u>. Refer to Appendix 2 (Mortuary Affairs) or, if not used, indicate the mortuary affairs activities applicable to the operation order or plan and policy for providing these affairs.

(9) (U) <u>Ammunition</u>. Refer to Appendix 6 (Nonnuclear Ammunition) or if not used, discuss any critical ammunition issues that may affect the ability of the force to accomplish the mission.

Page number
CLASSIFICATION

CLASSIFICATION

(10) (U) <u>Aviation Logistic Support</u>. Refer to Appendix 10 (Aviation Logistic Support) or Annex D (Logistics/Combat Service Support) of the aviation combat element operation order or plan. Critical aviation logistic and combat service support issues may be discussed if they affect the ability of the force to accomplish the mission.

(11) (U) <u>Operational Security Planning Guidance for Logistics</u>. Refer to Tab C (Operations Security) to Appendix 3 (Information Operations) to Annex C (Operations). Provide comprehensive operations security planning guidance for planning, preparing, and executing logistic and combat service support activities. At a minimum, address base, facility, installation, logistic stocks, physical, and line of communications security. Provide guidance to ensure that logistic and combat service support activities promote essential secrecy for operational intentions, capabilities that will be committed to specific missions, and current preparatory operational activities.

  b. (U) <u>Administration</u>. Include general administrative guidance to support logistic and combat service support operations for the basic operation order or plan. If reports are required, specify formats for preparation, time, methods, and classification of submission.

5. (U) <u>Command and Signal</u>

  a. (U) <u>Command Relationships</u>. Refer to Annex J (Command Relationships) for command relationships external to logistic units. Provide support relationships.

  b. (U) <u>Communications System</u>. Refer to Annex K (Combat Information Systems) for detailed communications and information systems requirements. Provide a general statement of the scope and type of communications required.

ACKNOWLEDGE RECEIPT

                                                Name
                                          Rank and Service
                                              Title

APPENDICES:

1–Petroleum, Oils, and Lubricants Supply
2–Mortuary Affairs
3–Sustainability Analysis
4–Mobility and Transportation
5–Civil Engineering Support Plan

                                              Page number
CLASSIFICATION

CLASSIFICATION

6–Nonnuclear Ammunition
7–Supply
8–Services
9–Health Services
10–Aviation Logistic Support (Normally provided in the aviation combat element plan or order.)
11–External Support
12–Maintenance
13–General Engineering

OFFICIAL:

s/ Name
Rank and Service
Title

APPENDIX D
LOGISTIC PLANNING CONSIDERATION FOR MCPP

Logistic or support planners may perceive the Marine Corps Planning Process (MCPP) as not being logistics-centric or only used by MAGTF-level staffs. Marine Corps Warfighting Publication 5-1 is written at the MAGTF level and most formal learning centers teach MCPP with maneuver centered scenarios which only require logistic estimates to support maneuver's COAs. Logistics exists to support maneuver, but only learning MCPP through a maneuver lens can result in logisticians with a passive focus on logistic estimates and support procedures. The MCPP is a universally effective analytical decision process that provides a better understanding of capabilities compared to those we are supporting and within the environment we will be operating in. Therefore, MCPP is malleable to the uniqueness of the organization using it and the mission, doctrine is not prescriptive because planning products cannot be predictive. The MCPP is the Marine Corps approach to providing collective understanding and enabling leaders to be proactive. The following discussion is intended to provide logistic planner's additional tools to apply to their situation and shape their planning efforts.

Problem Framing

Problem framing is understood to be the most important step of MCPP because no amount of planning can solve a problem that is insufficiently understood. However, it is important for planners to recognize they will never completely understand a complex problem and the planning process itself will continue to reveal new aspects of a problem. Planners should go into problem framing with the goal of understanding the purpose of their mission, the tasks they have to accomplish towards this purpose, and a better understanding of the environment they will operate in.

0. Discuss design concept with the commander
    a. Receive commander's orientation
    b. Define the environment
        1) Area of operations
        2) Area of interest
        3) Area of influence
        4) Information environment
        5) Culture
        6) Language
        7) Demographics
        8) Religion
        9) Geography
        10) Local economics
        11) Key actors
        12) Tendencies
        13) Relationships
        14) Security
        15) Climate
        16) Time

c. Define the problem
    1) Analyze HHQ mission
    2) Analyze HHQ intent
    3) Analyze HHQ order
    4) Analyze adversary
    5) Update changes to the friendly force
    6) Analyze the information environment
    7) Analyze the effects of terrain and weather on operations
    8) Determine troops and support available
    9) Analyze civil considerations (to include indigenous/local population)
    10) Differentiate between existing and desired conditions
    11) Analyze input from other commanders
    12) Determine the range of potential actions
d. Receive commander's initial intent
e. Receive commander's initial guidance

1. OPT-Problem Framing
2. Create situation update
    a. Staff estimates
    b. Functional estimates
    C. Estimates of supportability
3. Create IPB products **(direct the IPB)**
    a. Define the operational environment/battlespace environment
        1) Identify significant characteristics of the environment
        2) Identify the limits of the AO
        3) Establish the limits of the area of influence and the area of interest
        4) Evaluate existing databases and identify intelligence gaps
        5) Initiate collection of information required to complete IPB
    b. Describe environmental effects on operations/describe the battlespace effects
        1) Analyze the environment
        2) Describe the environmental effects on threat and friendly operations
        3) Courses of action/describe the battlespace effects on operations
        4) Adversary and friendly capabilities and courses of action
    c. Evaluate the threat/adversary
        1) Define threat/adversary
        2) Update or create threat/adversary models
        3) Identify threat/adversary capabilities
    d. Determine threat/adversary courses of action
        1) Identify the threat's/adversary's likely objectives and desired end state
        2) Identify the full set of courses of action available to the threat/adversary
        3) Evaluate and prioritize each course of action

4) Develop each course of action
5) Identify initial ISR requirements

4. Analyze PNA (physical network analysis)
    a. Locate infrastructure
        1) Airfields
        2) Ports
        3) Navigable rivers
        4) Navigable inland waterway
        5) Beach landing sites
        6) Roads
        7) Railways
        8) Tunnels
        9) Bridges
    b. Locate logistic infrastructure
        1) Water treatment
        2) Water storage facilities
        3) Petro storage
        4) Petro pipelines
        5) Medical facilities
        6) Warehousing
        7) Acreage
        8) Power generation facilities
        9) Power generation distribution

5. Incorporate HHQ mission
6. Incorporate HHQ commander's intent two levels up
7. Incorporate commander's initial intent
8. Incorporate commander's initial guidance
9. Conduct task analysis
    a. Specified
    b. Implied
    c. Essential
    d. Adjacent/supported
10. Develop assumptions
11. Determine limitations
    a. Restraints
    b. Constraints
12. Identify shortfalls
13. Conduct center of gravity analysis
    a. Friendly
    b. Supported unit

c. Enemy: Logistic planners can become creative with COG analysis. The enemy may not be the conventional enemy most planners use. A logistic unit may see the enemy as a friendly tank company that is trying to out consume what a direct support CLB can provide. This type of creative thinking enables a creative approach to how planners and commanders use their capabilities against the greatest threat to their mission.

14. Manage RFI process
    a. Create
    b. Submit
    c. Organize (info management)
    d. Disseminate responses
    e. Archive
    f. Brief answers as facts
    g. Follow up on unanswered RFIs
15. Recommend CCIRs
    a. FFIRs
    b. PIRs
    c. EEFIs
16. Recommend mission statement
17. Recommend COA development guidance
18. Manage planning product information
19. Establish red cell
20. Establish green cell

21. Determine essential elements of the plan to assess
    a. Integrate commander's end state into initial assessment planning
    b. Integrate essential tasks into the assessment plan as required
    c. Determine objectives

22. Analyze the information necessary to create the assessment plan
    a. Analyze IPB into assessment planning to establish baseline
    b. Analyze HHQ assessment annex
    c. Analyze HHQ mission
    d. Analyze HHQ objectives
    e. Analyze HHQ concept of operations

23. Develop the assessment plan
    a. Incorporate IPB into assessment planning to establish baseline
    b. Incorporate HHQ assessment annex
    c. Incorporate HHQ mission
    d. Incorporate HHQ objectives
    e. Incorporate HHQ concept of operations

24. Identify the capabilities of the 6 functions of logistics across friendly forces
    a. Analyze friendly mission
    b. Analyze friendly TO
    c. Analyze friendly TE

  d. Analyze friendly force organization
  e. Analyze logistic operations cycle
  f. Analyze potential locations of capabilities
25. **Correlate interrelationships of logistic functions across friendly forces**
  a. Analyze support requirements
  b. Determine logistic shortfalls
  c. Analyze shortfall sourcing options (external agencies)
  d. Analyze relationships of units performing the same function of logistics (across the MAGTF)
  e. Analyze the information (reporting) requirements for interrelationships across functions of logistics
  f. Analyze HHQ Concept of Logistics Support
26. **Identify capabilities of warfighting functions across friendly forces**
  a. Analyze friendly TO
  b. Analyze friendly TE
  c. Analyze friendly force organization
27. **Correlate interrelationships of warfighting functions across friendly forces**
  a. Analyze HHQ concept of fires support
  b. Analyze fires requirements
  c. Determine fires shortfalls
  d. Analyze fires shortfall sourcing options
  e. Analyze HHQ concept of intelligence support
  f. Analyze intelligence requirements
  g. Determine intelligence shortfalls
  h. Analyze intelligence shortfall sourcing options
  i. Analyze HHQ command and control plan
  j. Analyze command and control requirements
  k. Determine command and control shortfalls
  l. Analyze command and control sourcing options
  m. Analyze HHQ force protection plan
  n. Analyze force protection requirements
  o. Determine force protection shortfalls
  p. Analyze force protection sourcing options
  q. Analyze HHQ concept of operations
  r. Analyze maneuver requirements
  s. Determine maneuver shortfalls
  t. Analyze maneuver sourcing options
28. Develop recommended commander's COA development guidance
29. Draft warning order
30. Conduct problem framing brief
  a. Brief situation update
  b. Brief IPB and PNA

  c. Brief HHQ mission
  d. Brief HHQ commander's intent two levels up
  e. Brief commander's initial intent
  f. Brief commander's initial guidance
  g. Brief task analysis
  h. Brief assumptions
  i. Brief facts
  j. Brief RFIs
  k. Brief limitations
  l. Brief shortfalls
  m. Brief COG analyses
  n. Brief recommended CCIRs
  o. Brief proposed mission statement
  p. Brief recommended commander's COA development guidance
31. Receive commander's approval of mission statement
32. Receive commander's COA development guidance
33. Release warning order
  a. Update
  b. Issue
  c. Confirm receipt

COA Development

0. Discuss design concept with the commander
  a. Integrate commander's orientation
  b. Integrate commander's initial intent
  c. Integrate commander's initial guidance
1. **OPT-COA development**
2. Update situation
  a. Update staff estimates
  b. Update functional estimates
  c. Update estimates of supportability
3. Update intelligence preparation of the battlespace (IPB) products (**direct the IPB**)
  a. Refine the operational environment/battlespace environment
    1) Refine significant characteristics of the environment
    2) Refine the limits of the AO
    3) Refine the limits of the area of influence and the area of interest
    4) Refine existing databases and identify intelligence gaps
    5) Refine collection of information required to complete IPB
  b. Update environmental effects on operations/describe the battlespace effects
    1) Update the environment

2) Refine the environmental effects on threat and friendly operations
3) Refine the environmental effects on courses of action
4) Refine the battlespace effects on operations
5) Update adversary and friendly capabilities and courses of action
   c. Update the threat/adversary
      1) Update threat/adversary
      2) Update threat/adversary models
      3) Update threat/adversary capabilities
   d. Validate threat/adversary courses of action
      1) Validate the threat's/adversary's likely objectives and desired end state
      2) Validate the full set of courses of action available to the threat/adversary
      3) Validate evaluated and prioritized courses of action
      4) Validate each course of action
      5) Validate initial ISR requirements

4. Analyze PNA (physical network analysis)
   a. Refine PNA for relevant infrastructure
      1) Airfields
      2) Ports
      3) Navigable rivers
      4) Navigable inland waterway
      5) Beach landing sites
      6) Roads
      7) Railways
      8) Tunnels
      9) Bridges
   b. Refine PNA for relevant logistic infrastructure
      1) Water treatment
      2) Water storage facilities
      3) Petro storage
      4) Petro pipelines
      5) Medical facilities
      6) Warehousing
      7) Acreage
      8) Power generation facilities
      9) Power generation distribution

5. Update assumptions
   a. Answered RFIs become facts
   b. New assumptions captured and RFIs submitted
6. Incorporate Limitations
   a. Incorporate restraints
   b. Incorporate constraints

7. Update shortfalls
    a. Refine existing shortfalls
    b. Identify new shortfalls
8. Refine and Incorporate center of gravity analysis
    a. Friendly
    b. Supported unit
    c. Enemy
9. Update RFIs
    a. Update
    b. Submit
    c. Organize (info management)
    d. Disseminate responses
    e. Archive
    f. Brief answers as facts
    g. Follow up on unanswered RFIs
10. Refine CCIRs
    a. FFIRs
    b. PIRs
    c. EEFIs
11. Incorporate approved mission statement
12. Develop recommend COA wargaming guidance
13. Manage planning product information
14. Maintain red cell planning
15. Maintain green cell planning
16. Develop the assessment plan
    a. Integrate commander's end state into initial assessment planning
    b. Integrate essential tasks into the assessment plan as required
    c. Determine objectives tied to end state
    d. Determine tasks tied to objectives
    e. Determine measures of performance (MOPs) tied to tasks
    f. Determine measures of effectiveness (MOEs) tied to objectives
    g. Determine indicators tied to measures of effectiveness (MOEs)
    h. Incorporate IPB into assessment planning to establish baseline
    i. Incorporate HHQ assessment annex
    j. Incorporate HHQ mission
    k. Incorporate HHQ objectives
    l. Incorporate HHQ concept of operations
17. Identify the capabilities of the six functions of logistic across friendly forces
    a. Update friendly mission
    b. Update friendly TO
    c. Update friendly TE

       d. Update friendly force organization
       e. Update logistic operations cycle
       f. Update potential locations of capabilities

18. Correlate interrelationships of logistic functions across friendly forces
       a. Update support requirements
       b. Update logistic shortfalls (annex C and D)
       c. Update shortfall sourcing options (external agencies)
       d. Update relationships of units performing the same function of logistics (across the MAGTF)
       e. Update the information (reporting) requirements for interrelationships across functions of logistics
       f. Update HHQ Concept of Logistics Support

19. Identify capabilities of warfighting functions across friendly forces
       a. Update friendly TO
       b. Update friendly TE
       c. Update friendly force organization

20. Correlate interrelationships of warfighting functions across friendly forces
       a. Integrate HHQ concept of fires support into COAs
       b. Integrate fires requirements into COAs
       c. Integrate fires shortfalls into COAs
       d. Integrate fires shortfall sourcing options into COAs
       e. Integrate HHQ concept of intelligence support into COAs
       f. Integrate intelligence requirements into COAs
       g. Integrate intelligence shortfalls into COAs
       h. Integrate intelligence shortfall sourcing options into COAs
       i. Integrate HHQ command and control plan into COAs
       j. Integrate command and control requirements into COAs
       k. Integrate command and control shortfalls into COAs
       l. Integrate command and control sourcing options into COAs
       m. Integrate HHQ force protection plan into COAs
       n. Integrate force protection requirements into COAs
       o. Integrate force protection shortfalls into COAs
       p. Integrate force protection sourcing options into COAs
       q. Integrate HHQ concept of operations into COAs
       r. Integrate maneuver requirements into COAs
       s. Integrate maneuver shortfalls into COAs
       t. Integrate maneuver sourcing options into COAs

21. Establish task organization
       a. Determine structure of resources

22. Develop course of action
       a. Establish battlespace framework
       b. Determine array of forces

       c. Assign purpose

       d. Assign task

       e. Integrate actions across time and space

       f. Determine control measures

       g. Determine adversary

       h. Incorporate commander's intent

       i. Incorporate commander's guidance

23. Develop course of action graphic
24. Develop course of action narrative
25. Develop synchronization matrix
26. Prepare supporting concepts

       a. Develop functional concepts for each COA

       b. Develop supporting concepts for each COA

       c. Ensure actions are integrated for each COA

       d. Ensure actions are coordinated for each COA

27. Develop recommended commander's wargaming guidance
28. Develop recommended commander's evaluation criteria
29. Conduct course of action brief

       a. Brief HHQ mission

       b. Brief HHQ commander's intent two levels up

       c. Brief commander's initial intent

       d. Brief commander's initial guidance

       e. Brief COA graphic

       f. Brief COA narrative

       g. Brief refined facts

       h. Brief refined assumptions

       i. Brief outstanding/unanswered RFIs

       j. Brief possible adversary COAs

       k. Brief rational for each COA

       l. Brief recommended commander's wargaming guidance and evaluation criteria

30. Receive commander's refinements to COAs
31. Receive commander's COA wargaming guidance
32. Receive commander's evaluation criteria

MCTP 3-40B. Tactical-Level Logistics

<u>**COA Wargaming:**</u>

Wargaming offers challenges to logistic planners. The LCE may not use the traditional enemy to wargame against. Logistics, as a warfighting function, typically does not maneuver against the enemy. Therefore, creativity and critical thinking are required to stress developed COAs. Challenges within the environment, degraded road networks, interrupted supply nodes, unplanned GCE movements, etc. can provide commanders more realistic perspectives of their COAs than physical enemy activity.

0. Discuss design concept with the commander
    a. Integrate commander's wargaming guidance
    b. Integrate commander's initial intent
    c. Integrate commander's initial guidance
1. OPT-COA Wargaming
2. Update situation
    a. Update staff estimates
    b. Update functional estimates
    c. Update estimates of supportability
3. Update intelligence preparation of the battlespace (IPB) products (**direct the IPB**)
    a. Refine the operational environment/battlespace environment
        1) Refine significant characteristics of the environment
        2) Refine the limits of the AO
        3) Refine the limits of the area of influence and the area of interest
        4) Refine existing databases and identify intelligence gaps
        5) Refine collection of information required to complete IPB
    b. Update environmental effects on operations/describe the battlespace effects
        1) Update the environment
        2) Refine the environmental effects on threat and friendly operations
        3) Refine the environmental effects on courses of action
        4) Refine the battlespace efects on operations
        5) Update adversary and friendly capabilities and courses of action
    c. Update the threat/adversary
        1) Update threat/adversary
        2) Update threat/adversary models
        3) Update threat/adversary capabilities
    d. Validate threat/adversary courses of action
        1) Validate the threat's/adversary's likely objectives and desired end state
        2) Validate the full set of courses of action available to the threat/adversary
        3) Validate evaluated and prioritized courses of action
        4) Validate each course of action
        5) Validate initial ISR requirements

4. Analyze PNA
    a. Refine PNA for relevant infrastructure
        1) Airfields
        2) Ports
        3) Navigable rivers
        4) Navigable inland waterway
        5) Beach landing sites
        6) Roads
        7) Railways
        8) Tunnels
        9) Bridges
    b. Refine PNA for relevant logistic infrastructure
        1) Water treatment
        2) Water storage facilities
        3) Petro storage
        4) Petro pipelines
        5) Medical facilities
        6) Warehousing
        7) Acreage
        8) Power generation facilities
        9) Power generation distribution

5. Update assumptions
    a. Answered RFIs become facts
    b. New assumptions captured and RFIs submitted

6. Incorporate Limitations
    a. Incorporate restraints
    b. Incorporate constraints

7. Update shortfalls
    a. Refine existing shortfalls
    b. Identify new shortfalls

8. Refine and Incorporate center of gravity analysis
    a. Friendly
    b. Supported unit
    c. Enemy

9. Update RFIs
    a. Update
    b. Submit
    c. Organize (info management)
    d. Disseminate responses
    e. Archive

    f. Brief answers as facts
    g. Follow up on unanswered RFIs
10. Refine CCIRs
    a. FFIRs
    b. PIRs
    c. EEFIs
110. Incorporate approved mission statement
12. Develop recommended commander's COA comparison and decision guidance
13. Manage planning product information
14. **Develop the assessment plan** (as required)
    a. Refine objectives tied to end state
    b. Refine tasks tied to objectives
    c. Refine measures of performance (MOPs) tied to tasks
    d. Refine measures of effectiveness (MOEs) tied to objectives
    e. Refine indicators tied to measures of effectiveness (MOEs)
15. **Identify the capabilities of the six functions of logistics across friendly forces**
    a. Update friendly mission
    b. Update friendly TO
    c. Update friendly TE
    d. Update friendly force organization
    e. Update logistic operations cycle
    f. Update potential locations of capabilities
16. **Correlate interrelationships of logistic functions across friendly forces**
    a. Update support requirements based on wargame
    b. Update logistic shortfalls (annexes C and D)
    c. Update shortfall sourcing options (external agencies)
    d. Update relationships of units performing the same function of logistics (across the MAGTF)
    e. Update the information (reporting) requirements for interrelationships across functions of logistics
    f. Update HHQ Concept of Logistics Support
17. **Identify capabilities of warfighting functions across friendly forces**
    a. Update friendly TO
    b. Update friendly TE
    c. Update friendly force organization
18. **Correlate interrelationships of warfighting functions across friendly forces**
    a. Integrate HHQ concept of fires support (appendix 19 to annex C) into wargame
    b. Integrate fires requirements (appendix 19 to annex C) into wargame
    c. Integrate fires shortfalls into wargame
    d. Integrate fires shortfall sourcing options into wargame
    e. Integrate HHQ concept of intelligence support (annex B) into wargame
    f. Integrate intelligence requirements (annex B) into wargame

    g. Integrate intelligence shortfalls into wargame
    h. Integrate intelligence shortfall sourcing options into wargame
    i. Integrate HHQ command and control plan (annexes J, K, and U) into wargame
    j. Integrate command and control requirements (annexes J, K, and U) into wargame
    k. Integrate command and control shortfalls into wargame
    l. Integrate command and control sourcing options into wargame
    m. Integrate HHQ force protection plan (appendices 15 and 16 to annex C) into wargame
    n. Integrate force protection requirements (appendices 15 and 16 to annex C) into wargame
    o. Integrate force protection shortfalls into wargame
    p. Integrate force protection sourcing options into wargame
    q. Integrate HHQ concept of operations (annex C) into wargame
    r. Integrate maneuver requirements (annex C) into wargame
    s. Integrate maneuver shortfalls into wargame
    t. Integrate maneuver sourcing options into wargame

19. Develop a COA wargaming worksheet
    a. Include supported unit action
    b. Include LCE action
    c. Include enemy reaction
    d. Include host nation reaction
    e. Include supported unit counteraction
    f. Include LCE counteraction
    g. Include required assets
    h. Include approximate time
    i. Include decision points
    j. Include CCIRs
    k. Include remarks

20. Conduct COA wargame
    a. Determine supported unit action for each turn
    b. Determine LCE action for each turn
    c. Determine enemy reaction for each turn
        1) Implement Red Cell
    d. Determine host nation reaction for each turn
        1) Implement Green Cell
        2) Green cell discusses anticipated civilian population responses to friendly and adversary actions, reactions, and counteractions
    e. Determine supported unit counteraction for each turn
    f. Determine LCE counteraction for each turn
    g. Determine required assets
    h. Determine approximate time
    i. Determine decision points
    j. Determine CCIRs

k. Determine remarks

l. Evaluate each COA independently

m. Remain unbiased and avoid premature conclusions

n. Continue determination if each COA is suitable, feasible, acceptable, distinguishable, and complete

o. Record the advantages and disadvantages of each COA

p. Record issues and mitigations for risk, assumptions, and limitations

q. Record data based on commander's evaluation criteria for each COA

r. Keep to the established timeline of the war game

s. Identify possible branches and sequels for further planning

21. Develop a decision support template (graphical)

    a. Include decision points

    b. Include timeline

    c. Include NAIs

    d. Include TAIs

    e. Include restricted terrain

    f. Include avenues of approach

    g. Include lines of communication

22. Develop a decision support matrix (verbal)

    a. Include event numbers

    b. Include event

    c. Include NET

    d. Include NLT

    e. Include NAIs

    f. Include TAIs

    g. Include friendly action

23. Update functional and supporting concepts based on wargaming results (may include but not limited to the following)

    a. Risk assessment

    b. Casualty projections and limitations

    c. Personnel replacement requirements

    d. Projected enemy losses

    e. Enemy prisoner of war procedures

    f. Intelligence collection requirements and limitations

    j. Support (fires, information operations, logistics, aviation) strengths and limitations

    k. Projected assets and resource requirements

    l. Operational reach

    m. Projected allocation of mobility assets, lift, and sorties versus availability

    n. Requirement for prepositioning equipment and supplies

    o. Projected location of units and supplies for future operations

p. Projected location of the combat operations center and command post echelons

q. Command and control system's requirements

24. Develop updates to COAs based on wargaming results (may include but not limited to the following)
    a. Uncover new implied tasks
    b. Update task organization
    c. Update laydown locations
    d. Update COA GLOCs
    e. Update COA ALOCs
    f. Update COA SLOCs
    d. Update COA APOEs/APODs
    e. Update COA SPOEs/SPODs
    f. Update COA timelines
    g. Update synch matrix

25. Conduct COA wargame brief
    a. Brief HHQ mission
    b. Brief HHQ commander's intent two levels up
    c. Brief commander's initial intent
    d. Brief commander's initial guidance
    e. Brief COA overview (graphic and narrative)
    f. Brief refined facts
    g. Brief refined assumptions
    h. Brief outstanding/unanswered RFIs
    i. Wargame technique used
    j. Advantages of each COA
    k. Disadvantages of each COA
    l. Enemy COA situation templates
        1) Updated intelligence estimate
            <u>a</u> Terrain
            <u>b</u> Weather
            <u>c</u> Adversaries
            <u>d</u> Local population
        2) Wargamed responses of the population
    m. COA war game products and results
        1) COA war game worksheet
        2) Identification of any additional tasks
        3) Revised COA graphic and narrative
        4) Decision support template and matrix
        5) Revised synchronization matrix
        6) Branches and potential sequels

           7) Estimated time required for the operation
           8) Risk assessment
      n. Recommended commander's COA comparison and decision guidance

26. Receive commander's input
      a. Commander's approval to any changes to COAs
      b. Commander's COA comparison and decision guidance

COA Comparison and Decision

0. Discuss design concept with the commander
      a. Integrate commander's COA comparison guidance
      b. Integrate commander's initial intent
      c. Integrate commander's initial guidance

1. OPT-COA comparison and decision

2. Update situation
      a. Update staff estimates
      b. Update functional estimates
      c. Update estimates of supportability

3. Update intelligence preparation of the battlespace (IPB) products (**direct the IPB**)
      a. Refine the operational environment/battlespace environment
           1) Refine significant characteristics of the environment
           2) Refine the limits of the AO
           3) Refine the limits of the area of influence and the area of interest
           4) Refine existing databases and identify intelligence gaps
           5) Refine collection of information required to complete IPB
      b. Update environmental effects on operations/describe the battlespace effects
           1) Update the environment
           2) Refine the environmental effects on threat and friendly operations
           3) Refine the environmental effects on courses of action
           4) Refine the battlespace effects on operations
           5) Update adversary and friendly capabilities and courses of action
      c. Update the threat/adversary
           1) Update threat/adversary
           2) Update threat/adversary models
           3) Update threat/adversary capabilities
      d. Validate threat/adversary courses of action
           1) Validate the a threat's/adversary's likely objectives and desired end state
           2) Validate the full set of courses of action available to the threat/adversary
           3) Validate evaluated and prioritized courses of action
           4) Validate each course of action
           5) Validate initial ISR requirements

4. Analyze PNA
- a. Refine PNA for relevant infrastructure
    1) Airfields
    2) Ports
    3) Navigable rivers
    4) Navigable inland waterway
    5) Beach landing sites
    6) Roads
    7) Railways
    8) Tunnels
    9) Bridges
- b. Refine PNA for relevant logistic infrastructure
    1) Water treatment
    2) Water storage facilities
    3) Petro storage
    4) Petro pipelines
    5) Medical facilities
    6) Warehousing
    7) Acreage
    8) Power generation facilities
    9) Power generation distribution

5. Update assumptions
    - a. Answered RFIs become facts
6. Incorporate limitations
    - a. Incorporate restraints
    - b. Incorporate constraints
7. Update shortfalls
    - a. Refine existing shortfalls
    - b. Identify new shortfalls
8. Refine and Incorporate center of gravity analysis
    - a. Friendly
    - b. Supported unit
    - c. Enemy
9. Update RFIs
    - a. Update
    - b. Submit
    - c. Organize (info management)
    - d. Disseminate responses
    - e. Archive
    - f. Brief answers as facts
    - g. Follow up on unanswered RFIs

10. Refine CCIRs
    a. FFIRs
    b. PIRs
    c. EEFIs
11. Incorporate approved mission statement
12. Manage planning product information
13. **Identify the capabilities of the six functions of logistics across friendly forces** (as required)
    a. Update friendly mission
    b. Update friendly TO
    c. Update friendly TE
    d. Update friendly force organization
    e. Update logistic operations cycle
    f. Update potential locations of capabilities
14. **Correlate interrelationships of logistic functions across friendly forces**
    a. Update support requirements
    b. Update logistic shortfalls (annexes C and D)
    c. Update shortfall sourcing options (external agencies)
    d. Update relationships of units performing the same function of logistics (across the MAGTF)
    e. Update the information (reporting) requirements for interrelationships across functions of logistics
    f. Update HHQ Concept of Logistics Support
15. **Identify capabilities of warfighting functions across friendly forces**
    a. Update friendly TO
    b. Update friendly TE
    c. Update friendly force organization
16. **Correlate interrelationships of warfighting functions across friendly forces**
    a. Integrate HHQ concept of fires support into comparison and decision
    b. Integrate fires requirements into comparison and decision
    c. Integrate fires shortfalls into comparison and decision
    d. Integrate fires shortfall sourcing options into comparison and decision
    e. Integrate HHQ concept of intelligence support into comparison and decision
    f. Integrate intelligence requirements into comparison and decision
    g. Integrate intelligence shortfalls into comparison and decision
    h. Integrate intelligence shortfall sourcing options into comparison and decision
    i. Integrate HHQ command and control plan into comparison and decision
    j. Integrate command and control requirements into comparison and decision
    k. Integrate command and control shortfalls into comparison and decision
    l. Integrate command and control sourcing options into wargame
    m. Integrate HHQ force protection plan into comparison and decision
    n. Integrate force protection requirements into comparison and decision
    o. Integrate force protection shortfalls into comparison and decision

      p. Integrate force protection sourcing options into comparison and decision
      q. Integrate HHQ concept of operations into comparison and decision
      r. Integrate maneuver requirements into comparison and decision
      s. Integrate maneuver shortfalls into comparison and decision
      t. Integrate maneuver sourcing options into comparison and decision
17. Conduct COA comparison
      a. Evaluate COAs independently
          1) Apply evaluation criteria
          2) Apply estimates of supportability
          3) Apply weight to criteria
      b. Decide comparison method
          1) Qualitative comparison technique
          2) Quantitative comparison technique
          3) Hybrid/blended comparison technique
      c. Compare COAs
          1) Create comparison and decision matrix
          2) Display weight of criteria
      d. Determine COA to recommend
18. Conduct COA comparison and decision brief
      a. Brief HHQ mission
      b. Brief HHQ commander's intent two levels up
      c. Brief commander's initial intent
      d. Brief commander's initial guidance
      e. Brief COA overview (graphic and narrative)
      f. Brief refined facts
      g. Brief outstanding assumptions
      h. Brief outstanding/unanswered RFIs
      i. Brief comparison method used
      j. Evaluation results for each COA
      k. Comparison results between COAs
      l. Completed comparison and decision matrix
19. Receive commander's input
      a. Commander's approval on a COA
      b. Commander's direction to modify a COA
      c. Commander's direction to merge two or more COAs
      d. Commander's direction to discard all COAs
20. Refine CONOPS
21. Update warning order
      a. Issue
      b. Confirm receipt

Orders Development

A written order is essential to clearly communicate critical information and provide a common understanding of the unit's problem and goals. Staff should not only focus on "proper" formatting, although important, but on putting critical thought into what must be articulated to properly execute the order. Higher headquarters information must not be regurgitated without purpose and key points developed during the planning process must be written concisely and clearly to be relevant for operations.

A common problem for the LCE is that the MCWP 5-1, or any other reference, does not direct how LCEs write annex C or annex D, it is dependent of commanders and the situation. Often in an LCE, annex C is written as logistic operations and annex D as internal logistics, but there are other methods that can be used. The key point is to ensure a discussion takes place between the commander and his staff about how operational logistics is written within the order and the decision is communicated to HHQ, supported, and adjacent units.

The ultimate test of an order's quality is when an officer or noncommissioned officer, not exposed to the products planning processes, understands the final product and can apply it.

0. Base order
    a. Does the order succinctly provide information relevant to providing SA?
    b. Is battlespace presented accurately? Are maps provided?
    c. Is enemy described in a way relevant to combat support?
    d. Did the order get creative and discuss friendly forces COG in relation to logistics?
    e. Were attachments/detachments accurately described?
    f. Is the mission and execution communicated clearly?
        1) Did either get creative and identify a problem to the reader?
        2) How well does CONOPS communicate summary to reader?
        3) Does CONOPS accurately capture COA chosen?
        4) Are tasks clear with purpose and necessary for CONOPS?
    g. The focus is not the mechanics of writing or nuances of formatting.
        1) Thoughtful content and clear communication of how to execute a synthesized plan is the goal of an order.
        2) Ask yourself, "Can I make a plan and execute based off of this order?"
        3) Annex B. Intelligence: Not all inclusive, but typical key appendices for LCEs appendix 9.
            (a) Captured adversary documentation and evacuation procedures for personal items of intelligence value.
            (b) Indicates items/provides guidance for items that should not be collected. Receive commander's orientation.
1. Appendix 11. Intelligence Estimate
    a. Tab A. Tactical Study of Terrain
        1) Cover, concealment, natural obstacles, observation/fields of fire, avenues of approach/escape.

b. Tab B. Beach Studies
   c. Condition of sand (loose/packed)
      1) Best position for amphibious bulk liquid transfer system (ABLATS)
      2) Terrain that challenges of establishing ABLATS
      3) Location of largest flat/level area from beach to establish fuel farm
   d. Tab C. Climatology
      1) Visibility, precipitation, temperature impacts to operations.
   e. Tab D. Airfields
      1) Aircraft supportability (weight, runway length), adequate marshalling area access.
   f. Tab E. HLZ/Drop Zones
      1) Overhead obstacles
      2) Route access
   g. Tab F. Port Studies
   h. Water depth, warehouse access, stevedore/MHE availability
2. Appendix 13. Intelligence Collection Plan
   a. Tab C. CLR NAIs
      1) Tied to enemy activity should be an area that the CLR anticipates the enemy to have activity in/around.
      2) Clear, appropriate size, NAI
      3) Possible Nondoctrinal approach, have they tied NAI to GCE support?
3. Annex C. Operations
   a. Appendix 2. CBRN
   b. Pass through what is relevant. The whole Appendix isn't just copy and paste.
   c. Appendix 3. Information Operations
      1) Anchored with ability to impact friendly mission accomplishment, influences on adversary's decisionmaking process, and importance of swaying otherwise neutral parties.
      2) Is this appendix developed to be relevant based upon MEB order?
      3) Tab C is developed to be relevant based upon MEB order.
   d. Appendix 5. Evasion and Recovery Operations
      1) Ensure this appendix and Appendix 11 gets consolidated and is not just repeat information.
   e. Appendix 6. Rules of Engagement
      1) Pass through from HHQ, but ensure right to defend yourself, return of hostile fire, use of minimum force for hostile but unarmed people, property seizure, civilian detention are simple and easy to remember.
   f. Appendix 13. Explosive Ordnance Disposal
      1) ID how EOD is organized within the MEB.
      2) If CLR is EOD Ops must be developed by the CLR.
   g. Appendix 16. Rear Area Operations.
      1) Does CLR have tasks in RAO?
      2) Have appropriate taskings been given in the areas of: security, sustainment, infrastructure development, HNS, and intelligence?

h. Appendix 18. Operations Overlay.
   1) Appropriate use of ops terms and graphics, color-coded.
   2) LCE ops terms and graphics included. Not just GCE.
   3) PNA overlay.
i. Appendix 19. Fire Support
   1) Appendix 19 from the MEB is passed through with thought put into what critical information is relevant to their subordinates. Not just copy and paste.
   2) Tab B. Artillery Support Plan - Build a tab to hang the support plan.
   3) Enclosure 1. Artillery Target List – Included without targets outside MEB AO.
      (a) Enclosure 2. Artillery Synchronization Matrix – Pass through info.
      (b) Enclosure 3. Artillery Target Overlay – Pass through information.
   4) Tab F. Fire Support Coordination Plan – Make basic document to hang enclosures.
      (a) Enclosure 2 HPT TSS AGM - Pass through info from HHQ.
4. Annex D. LogisticsCombat Service Support
   a. Appendix 1. Petroleum, Oils, and Lubricants.
      1) Does this appendix and tabs A, B, C, D present a clear picture of how POL storage and distribution will be conducted?
   b. Appendix 2. Mortuary Affairs
      1) Is retrieval, identification, transportation, and storage addressed?
   c. Appendix 5. Civil Engineering Support Plan.
      1) Are naval construction forces addressed and their role communicated?
      2) How are MEB engineers depicted in this appendix?
      3) Does the order spend too much time communicating civil engineer support that is not realistic or possible?
   d. Appendix 7. Supply
      1) Is there a priority of transportation by phase?
      2) Does the order present a clear picture on classes of supply and LCE responsibility?
      3) Is the process for procurement (order/delivery) of the classes of supply communicated?
   e. Appendix 11. External Support
      1) What is the process to identify the requirement for external support?
   f. Appendix 12. Maintenance
      1) Where are the intermediate and depot levels of maintenance available? Briefly and generally, what is the process to induct equipment into the maintenance cycle?
   g. Appendix 13. General Engineering
      1) Is there a priority of engineers by phase?
      2) Is there a concept for employment of engineers?
      3) Is a process communicated for creating an engineer project list?
5. Annex E. Personnel
   a. Has there been thought put into transportation responsibilities of EPW, internees, or other detained persons?
6. Annex F. Public Affairs

  a. Is the appropriate information from MEB being passed on to subordinate units?
7. Annex G. Civil-Military Operations
  a. Is the appropriate information from MEB being passed on to subordinate units?
8. Annex J. Command Relationships
  a. Has a command relationship diagram been completed depicting what is being communicated within the order.
  b. How is EOD being depicted? How does that compare to annex C?
9. Annex P. Host-Nation Support
  a. What HNS is being relied on?
  b. If relying on HNS for shortfalls is that communicated here?
10. Annex Q. Medical Services
  a. Is this annex depicted as a standalone annex? Did the order incorporate medical services in other annexes, such as annex C?
11. Annex V. Interagency Coordination
  a. Does the order identify other organizations that may be encountered within the AO?
  b. Is the information presented in a way that may help subordinates gain SA on other agencies?
12. Annex W. Aviation Operations
  a. Does the order depict information that is relevant to subordinate units?
  b. Are processes for requesting air support included?
  c. Does the order only copy and paste this from MEB order or does it involve some thought?

GLOSSARY

Section I. Abbreviations and Acronyms

AAA	arrival and assembly area
AACG	arrival airfield control group
AAOG	arrival and assembly operations group
ACB	amphibious construction battalion
ACE	aviation combat element
ACSA	acquisition and cross-servicing agreement
ADAL	authorized dental allowance list
AFOE	assault follow-on echelon
ALD	aviation logistics division
ALOC	adminstration and logistic operations center
AMAL	authorized medical allowance list
AMC	Air Mobility Command
AO	area of operations
APOD	aerial port of debarkation
APOE	aerial port of embarkation
ASP	ammunition supply point
ATF	amphibious task force
AVCAL	aviation consolidated allowance list
AVLOG	aviation logistics
BSA	beach support area
C2	command and control
CAB	combat assualt battalion
CBMU	construction battalion maintenance unit
CE	command element
CEB	combat engineer battalion
CJCSM	Chairman of the Joint Chiefs of Staff manual
CJSI	Chairman of the Joint Chiefs of Staff instruction
CLB	combat logistics battalion
CLC	combat logistics company
CLR	combat logistics regiment
CMC	Commandant of the Marine Corps
COA	course of action
COC	combat operations center
COG	center of gravity
COMMARFOR	commander, Marine Corps forces
CONUS	continental United States
COSAL	coordinated ship-station allowance list

CRTS	casualty receiving and treatment ship
CSP	contingency support package
CSS	combat service support
CSSA	combat service support area
DACG	departure airfield control group
DC, I&L	Deputy Commandant for Installations and Logistics
DOD	Department of Defense
DODD	Department of Defense directive
EAF	expeditionary airfield
ECP	expeditionary contracting platoon
EOD	explosive ordnance disposal
EPW	enemy prisoner of war
ESB	engineer support battalion
FARP	forward arming and refueling point
FIE	fly-in echelon
FISP	fly-in support package
FOB	forward operating base
FOO	field ordering officer
FSR	first strike ration
G-1	assistant chief of staff, personnel
G-2	assistant chief of staff, intelligence
G-3	assistant chief of staff, operations
G-4	assistant chief of staff, logistics
G-5	assistant chief of staff, plans
G-6	assistant chief of staff, communications system
GCCS	Global Command and Control System
GCE	ground combat element
GCPC	Government-Wide Commercial Purchase Card
GCSS-MC	Global Combat Support System-Marine Corps
GSORTS	Global Status of Resources and Training
GS-R	general support-reinforcing
HHQ	higher headquarters
HN	host nation
HNS	host-nation support
HQ	headquarters
HQMC	Headquarters, United States Marine Corps
H&S	headquarters and service

HSS	health service support
HST	helicopter support team
IMA	intermediate maintenance activity
IMRL	individual material readiness list
IPB	intelligence preparation of the battlespace
J-4	logistics directorate of a joint staff
JDDOC	joint deployment and distribution operations center
JFC	joint force commander
JLOTS	joint logistics over-the-shore
JOPES	Joint Operation Planning and Execution System
JP	joint publication
JTF	joint task force
LCE	logistics combat element
LF	landing force
LFOC	landing force operations center
LFSP	landing force support party
LOC	lines of communication
LOTS	logistics over-the-shore
LRP	food packet, long range patrol
MAARS II	Marine Corps Ammunition Accounting and Reporting System II
MACG	Marine air control group
MAG	Marine aircraft group
MAGTF	Marine air-ground task force
MALS	Marine aviation logistics squadron
MALSP	Marine aviation logistics support program
MARCORLOGCOM	Marine Corps Logistics Command
MARCORSYSCOM	Marine Corps Systems Command
MARDIV	Marine division
MARFOR	Marine Corps forces
MARFORCOM	Marine Forces Command
MARFORPAC	Marine Forces Pacific
MARFORRES	Marine Forces Reserve
MAW	Marine aircraft wing
MCA	movement control agency
MCC	movement control center
MCCS	Marine Corps Community Services
MCDP	Marine Corps doctrinal publication
MCICOM	Marine Corps Installations Command
MCPP	Marine Corps Planning Process

MCPP-N	Marine Corps Prepositioning Program-Norway
MCRP	Marine Corps reference publication
MCT	maintenance contact team
MCW	meal, cold weather
MCWP	Marine Corps warfighting publication
MCX	Marine Corps Exchange
MDDOC	Marine air-ground task force deployment and distribution operations center
MEB	Marine expeditionary brigade
MEDLOGCO	medical logistics company
MEF	Marine expeditionary force
METT-T	mission, enemy, terrain and weather, troops and support available—time available
MEU	Marine expeditionary unit
MHE	materials handling equipment
MLG	Marine logistics group
MMCC	Marine air-ground task force movement control center
MMDC	MAGTF materiel distribution center
MORE	modular operational ration enhancement
MOS	military occupational specialty
MPF	maritime prepositioning force
MPS	maritime prepositioning ship
MPSRON	maritime prepositioning ships squadron
MRE	meal, ready to eat
MSC	major subordinate command
MSE	major subordinate element
MST	maintenance support team
MWHS	Marine wing headquarters squadron
MWR	morale, welfare, and recreation
MWSS	Marine wing support squadron
N-4	Navy component logistics staff officer
NATO	North Atlantic Treaty Organization
NBG	naval beach group
NCF	naval construction force
NCG	naval construction group
NCR	naval construction regiment
NDP	naval doctrine publication
NLI	Naval Logistics Integration
NMCB	naval mobile construction battalion
NSOR	nutritional standards for operational rations
NTTP	Navy tactics, techniques, and procedures

MCTP 3-40B. Tactical-Level Logistics

OCS	operational contract support
OIS-MC	Ordnance Informations System-Marine Corps
OPCON	operational control
OPLAN	operation plan
OPORD	operation order
OPT	operational planning team
PNA	physical network analysis
POD	port of debarkation
POE	port of embarkation
POR	packaged operational ration
RSupply	relational supply
S-1	personnel officer
S-2	intelligence officer
S-3	operations officer
S-4	logistics officer
S-6	communications systems officer
SDDC	Surface Deployment and Distribution Command
SECNAVINST	Secretary of the Navy instruction
SERMIS	support equipment resources management information system
SOFA	status-of-forces agreement
SOP	standing operating procedure
SPMAGTF	special purpose Marine air-ground task force
SPOD	seaport of debarkation
SPOE	seaport of embarkation
STS	ship-to-shore
TACC	tactical air control center (Navy)
TACLOG	tactical-logistical
TACON	tactical control
T-AVB	aviation logistics support ship (MSC)
TBA	table of basic allowance
TC-AIMS II	Transportation Coordinator's Automated Information for Movement System II
TE	table of equipment
T/M/S	type/model/series
TO	table of organization
TSB	transportation support battalion
UCT	underwater construction team
UGR	unitized group ration

UGR-A	unitized group ration-A
UGR-B	unitized group ration-B
UGR-H&S	unitized group ration-heat and serve
UGR-M	unitized group ration-M
UHT	ultra high temperature
UMCC	unit movement control center
USTRANSCOM	United States Transportation Command
VTOL	vertical takeoff and landing

Section II. Terms and Definitions

A

airfield—An area prepared for the accommodation (including any buildings, installations, and equipment), landing, and takeoff of aircraft. (JP 1-02)

allocation— Distribution of limited forces and resources for employment among competing requirements. (JP-1-02)

amphibious assault—A type of amphibious operation that involves establishing a force on a hostile or potentially hostile shore. (JP 1-02)

amphibious operation—A military operation launched from the sea by an amphibious force to conduct landing force operations within littorals. (JP 1-02)

area of responsibility—The geographical area associated with a combatant command within which a geographic combatant commander has authority to plan and conduct operations. (JP 1-02)

assign—1. To place units or personnel in an organization where such placement is relatively permanent, and/or where such organization controls and administers the units or personnel for the primary function, or greater portion of the functions, of the unit or personnel. 2. To detail individuals to specific duties or functions where such duties or functions are primary and/or relatively permanent. (JP 1-02)

B

basic load—The quantity of supplies required to be on hand within, and which can be moved by, a unit or formation, expressed according to the wartime organization of the unit or formation and maintained at the prescribed levels. (JP 1-02)

beach party—The Navy component of the landing force support party under the tactical control of the landing force support party commander. (JP 1-02)

beach support area—In amphibious operations, the area to the rear of a landing force or elements thereof, that contains the facilities for the unloading of troops and materiel and the support of the forces ashore. (JP-1-02)

C

campaign—A series of related military operations aimed at achieving strategic and operational objectives within a given time and space. (JP 1-02)

casualty—Any person who is lost to the organization by having been declared dead, duty status-whereabouts unknown, missing, ill, or injured. (JP 1-02)

casualty collection—The assembly of casualties at collection and treatment sites. It includes protection from further injury while awaiting evacuation to the next level of care. Planning for casualty collection points must include site selection and manning. (MCRP 5-12C)

casualty evacuation—(See JP 1-02 for core definition. Marine Corps amplification follows.) The movement of the sick, wounded, or injured. It begins at the point of injury or the onset of disease. It includes movement both to and between medical treatment facilities. All units have an evacuation capability. Any vehicle may be used to evacuate casualties. If a medical vehicle is not used it should be replaced with one at the first opportunity. Similarly, aeromedical evacuation should replace surface evacuation at the first opportunity. (MCRP 5-12C)

classes of supply—(See JP 1-02 for core definition. Marine Corps amplification follows.) The ten categories into which supplies are grouped in order to facilitate supply management and planning.
  a. Class I—Subsistence, which includes gratuitous health and welfare items and rations.
  b. Class II—Clothing, individual equipment, tentage, organizational tool sets and tool kits, hand tools, administrative and housekeeping supplies, and equipment.
  c. Class III—Petroleum, oils, and lubricants, which consists of petroleum fuels, lubricants, hydraulic and insulating oils, liquid and compressed gases, bulk chemical products, coolants, deicing and antifreeze compounds, preservatives together with components and additives of such products, and coal.
  d. Class IV—Construction, which includes all construction material; installed equipment; and all fortification, barrier, and bridging materials
  e. Class V—Ammunition of all types, which includes, but is not limited to, chemical, radiological, special weapons, bombs, explosives, mines, detonators, pyrotechnics, missiles, rockets, propellants, and fuzes.
    Class V(A) Aviation ammunition
    Class V(W) Ground ammunition
  f. Class VI—Personal demand items or nonmilitary sales items.
  g. Class VII—Major end items, which are the combination of end products assembled and configured in their intended form and ready for use (e.g., launchers, tanks, mobile machine shops, vehicles).
  h. Class VIII—Medical/dental material, which includes medical-unique repair parts, blood and blood products, and medical and dental material.
  i. Class IX—Repair parts (less Class VIII), including components, kits, assemblies, and subassemblies (reparable and nonreparable), required for maintenance support of all equipment.
  j. Class X—Material to support nonmilitary requirements and programs that are not included in classes I through IX. For example, materials needed for agricultural and economic development. (MCRP 5-12C)

combatant commander—A commander of one of the unified or specified combatant commands established by the President. Also called **CCDR.** (JP 1-02)

combat logistics battalion—The task-organized logistics combat element of the Marine expeditionary unit. Personnel and equipment are assigned from the permanent battalions of the

Marine logistics group. As required, it may be augmented by combat service support assets from the Marine division or Marine aircraft wing. Also called **CLB.** (MCRP 5-12C)

combat power—The total means of destructive and/or disruptive force which a military unit/formation can apply against the opponent at a given time. (JP 1-02)

combat service support—The essential capabilities, functions, activities, and tasks necessary to sustain all elements of operating forces in theater at all levels of war. Also called **CSS**. (JP 1-02)

combat service support area—(See JP 1-02 for core definition. Marine Corps amplification follows.) The primary combat service support installation established to support MAGTF operations ashore. Normally located near a beach, port, and/or an airfield, it usually contains the command post of the logistics combat element commander and supports other combat service support installations. Also called **CSSA.** (MCRP 5-12C)

command and control—(See JP 1-02 for core definition. Marine Corps amplification follows.) The means by which a commander recognizes what needs to be done and sees to it that appropriate actions are taken. Command and control is one of the six warfighting functions. Also called **C2**. (MCRP 5-12C)

concept of logistic support—A verbal or graphic statement, in a broad outline, of how a commander intends to support and integrate with a concept of operations in an operation or campaign. (JP 1-02)

concept of operations—A verbal or graphic statement that clearly and concisely expresses what the joint force commander intends to accomplish and how it will be done using available resources. Also called CONOPS. (JP 1-02)

contingency plan—A plan for major contingencies that can reasonably be anticipated in the principal geographic subareas of the command. (JP 1-02)

contracting—The purchasing, renting, leasing, or otherwise obtaining supplies or services from nonfederal sources. Contracting includes description (but not determination) of supplies and services required, selection and solicitation of sources, preparation and award of contracts, and all phases of contract administration. It does not include making grants or cooperative agreements. (Federal Acquisition Regulation)

contracting officer—A person with the authority to enter into, administer, and/or terminate contracts and make related determinations and findings. (Federal Acquisition Regulation) A Service member or Department of Defense civilian with the legal authority to enter into, administer, modify, and/or terminate contracts. (JP 1-02)

control—Authority that may be less than full command exercised by a commander over part of the activities of subordinate or other organizations. (This is part one of a four part definition.) (JP 1-02)

coordination—The action necessary to ensure adequately integrated relationships between separate organizations located in the same area. Coordination may include such matters as fire support, emergency defense measures, area intelligence, and other situations in which coordination is considered necessary. (MCRP 5-12C)

countermobility operations—The construction of obstacles and emplacement of minefields to delay, disrupt, and destroy the enemy by reinforcement of the terrain. (JP 1-02)

crisis action planning—The Adaptive Planning and Execution system process involving the time-sensitive development of joint operation plans and operations orders for the deployment, employment and sustainment of assigned and allocated forces and resources in response to an imminent crisis. (JP 1-02)

cross-servicing—A subset of common-user logistics in which a function is performed by one Military Service in support of another Service and for which reimbursement is required from the Service receiving support. (JP 1-02)

D

depot—1. **supply**—An activity for the receipt, classification, storage, accounting, issue, maintenance, procurement, manufacture, assembly, research, salvage, or disposal of material. 2. **personnel**—An activity for the reception, processing, training, assignment, and forwarding of personnel replacements. (JP 1-02)

depot-level maintenance— Maintenance actions taken on materiel or software involving the inspection, repair, overhaul, or the modification or reclamation (as necessary) of weapons systems, equipment end items, parts, components, assemblies, and sub-assemblies that are beyond field-level maintenance capabilities. (Proposed for inclusion in MCRP 5-12C)

distribution—1. The arrangement of troops for any purpose, such as a battle, march, or maneuver. 2. A planned pattern of projectiles about a point. 3. A planned spread of fire to cover a desired frontage or depth. 4. An official delivery of anything, such as orders or supplies. 5. The operational process of synchronizing all elements of the logistic system to deliver the "right things" to the "right place" at the "right time" to support the geographic combatant commander. **6.** The process of assigning military personnel to activities, units, or billets. (JP 1-02)

distribution point—A point at which supplies and/or ammunition, obtained from supporting supply points by a division or other unit, are broken down for distribution to subordinate units. (JP 1-02)

distribution system—That complex of facilities, installations, methods, and procedures designed to receive, store, maintain, distribute, and control the flow of military materiel between the point of receipt into the military system and the point of issue to using activities and units. (JP 1-02)

directive authority for logistics—Combatant commander authority to issue directives to subordinate commanders to ensure the effective execution of approved operation plans, optimize the use or reallocation of available resources, and prevent or eliminate redundant facilities and/or overlapping functions among the Service component commands. Also called **DAFL**. (JP 1-02)

E

embarkation—The process of putting personnel and/or vehicles and their associated stores and equipment into ships and/or aircraft. (JP 1-02)

engineer reconnaissance—The gathering of specific, detailed, technical information required by supporting engineer forces in order to prepare for and accomplish assigned missions. (MCRP 5-12C)

evacuation—The clearance of personnel, animals, or materiel from a given locality. The controlled process of collecting, classifying, and shipping unserviceable or abandoned materiel, United States or foreign, to appropriate reclamation, maintenance, technical intelligence, or disposal facilities. (Part two and three of a four-part definition.) (JP 1-02)

expeditionary airfield—A prefabricated and fully portable airfield. The effort and assets (e.g., materiel, engineer support, operational guidance, security) required for the installation/operation of an expeditionary airfield can require the participation/support of all elements of the Marine air-ground task force. (NTRP 1-02)

explosive ordnance—All munitions containing explosives, nuclear fission or fusion materials and biological and chemical agents. (JP 1-02)

explosive ordnance disposal—The detection, identification, on-site evaluation, rendering safe, recovery, and final disposal of unexploded explosive ordnance. (JP 1-02)

F

firepower—The amount of fire which may be delivered by a position, unit, or weapon system. (JP 1-02)

floating dump—Emergency supplies preloaded in landing craft, amphibious vehicles, or in landing ships that are located in the vicinity of the appropriate control officer who directs their landing as requested by the troop commander concerned. (JP 1-02)

forward arming and refueling point—A temporary facility, organized, equipped, and deployed to provide fuel and ammunition necessary for the employment of aviation maneuver units in combat. (JP 1-02)

G

general engineering—(See JP 1-02 for core definition. Marine Corps amplification follows.) Intensive effort by engineer units that involves high standards of design and construction as well as detailed planning and preparation. It is that wide range of tasks in rear areas that serves to sustain forward combat operations. (MCRP 5-12C)

Global Command and Control System—A deployable command and control system supporting forces for joint and multinational operations across the range of military operations with compatible, interoperable, and integrated communications systems. Also called **GCCS**. (JP 1-02)

H

horizontal and vertical construction—Deliberate engineering projects that normally involve time, manpower, material, and equipment-intensive tasks. These tasks usually relate to survivability and sustainability efforts. (MCRP 5-12C)

J

Joint Operation Planning and Execution System—An Adaptive Planning and Execution system technology. Also called **JOPES**. (JP 1-02)

L

landing force—A Marine Corps or Army task organization, which is part of the amphibious force, formed to conduct amphibious operations. Also called LF. (JP 1-02)

landing zone support area—A forward support installation that provides minimum essential support to the air assault forces of the Marine air-ground task force. It can expand into a combat service support area but it is most often a short-term installation with limited capabilities, normally containing dumps for rations, fuel, ammunition, and water only; maintenance is limited to contact teams and/or support teams. (MCRP 5-12C)

line of communications—A route, either land, water, and/or air, that connects an operating military force with a base of operations and along which supplies and military forces move. (JP 1-02)

logistics—(See JP 1-02 for core definition. Marine Corps amplification follows.) **1.** The science of planning and executing the movement and support of forces. 2. All activities required to move and sustain military forces. Logistics is one of the six warfighting functions. (MCRP 5-12C)

logistics combat element—The core element of a Marine air-ground task force (MAGTF) that is task-organized to provide the combat service support necessary to accomplish the MAGTF's mission. The logistics combat element varies in size from a small detachment to one or more Marine logistics groups. It provides supply, maintenance, transportation, general engineering, health services, and a variety of other services to the MAGTF. In a joint or multinational environment, it may also contain other Service or multinational forces assigned or attached to the MAGTF. The logistics combat element itself is not a formal command. Also called **LCE**. (MCRP 5-12C)

logistics support—Support that encompasses the logistics services, materiel, and transportation required to support the continental United States-based and worldwide deployed forces. (JP 1-02)

M

main supply route—The route or routes designated within an operational area upon which the bulk of traffic flows in support of military operations. (JP 1-02)

maintenance—1. All action, including inspection, testing, servicing, classification as to serviceability, repair, rebuilding, and reclamation, taken to retain materiel in a serviceable condition or to restore it to serviceability. 2. All supply and repair action taken to keep a force in condition to carry out its mission. 3. The routine recurring work required to keep a facility in such condition that it may be continuously used, at its original or designed capacity and efficiency for its intended purpose. (JP 1-02)

Marine aircraft group—Administratively and tactically structured by aircraft category as being either a helicopter group or a fixed-wing group. Marine aircraft groups may also be formed for specific missions or unique organizational/geographic considerations. Each Marine aircraft group has a headquarters and maintenance squadron. With a source of supply, the Marine aircraft group is the smallest aviation unit capable of self-sustaining, independent operations. Also called **MAG**. (MCRP 5-12C)

Marine aircraft wing—The highest level aviation command in the Fleet Marine Force. The MAW is task-organized to provide a flexible and balanced air combat organization capable of the full range of combat air operations in a variety of areas without the requirement of prepositioned support, control, and logistic facilities. Only the wing has the inherent capability of performing all six aviation functions. Also called **MAW**. (MCRP 5-12C)

Marine air-ground task force—The Marine Corps' principal organization for all missions across a range of military operations, composed of forces task-organized under a single commander capable of responding rapidly to a contingency anywhere in the world. The types of forces in the Marine air-ground task force (MAGTF) are functionally grouped into four core elements: a command element, an aviation combat element, a ground combat element, and a logistics combat element. The four core elements are categories of forces, not formal commands. The basic structure of the MAGTF never varies, though the number, size, and type of Marine Corps units comprising each of its four elements will always be mission dependent. The flexibility of the organizational structure allows for one or more subordinate MAGTFs to be assigned. In a joint or multinational environment, other Service or multinational forces may be assigned or attached. Also called **MAGTF**. (MCRP 5-12C)

Marine division—A ground force of combat and combat support units organized and equipped primarily for amphibious operations. It consists of three infantry regiments, an artillery regiment, and separate combat support battalions. Subordinate units can be organized into effective forces of combined arms based upon the infantry regiment, infantry battalion, or tank battalion. One or more divisions form the ground combat element of the Marine expeditionary force. To perform its combat role, it requires air defense and aviation support from a Marine aircraft wing and service support from a Marine logistics group. Also called **MARDIV**. (MCRP5-12C)

Marine expeditionary brigade—A Marine air-ground task force (MAGTF) that is constructed around an infantry regiment reinforced, a composite Marine aircraft group, and a combat logistics regiment. The Marine expeditionary brigade (MEB), commanded by a general officer, is task-organized to meet the requirements of a specific situation. It can function as part of a joint task force, as the lead echelon of the Marine expeditionary force (MEF), or alone. It varies in size and composition and is larger than a Marine expeditionary unit but smaller than a MEF. The MEB is capable of conducting missions across the range of military operations. In a joint or multinational environment, it may also contain other Service or multinational forces assigned or attached to the MAGTF. Also called **MEB.** (MCRP 5-12C)

Marine expeditionary force—The largest Marine air-ground task force (MAGTF) and the Marine Corps' principal warfighting organization, particularly for larger crises or contingencies. It is task-organized around a permanent command element and normally contains one or more Marine divisions, Marine aircraft wings, and Marine logistics groups. The Marine expeditionary force is capable of missions across a range of military operations, including amphibious assault and sustained operations ashore in any environment. It can operate from a sea base, a land base, or both. In a joint or multinational environment, it may also contain other Service or multinational forces assigned or attached to the MAGTF. Also called **MEF**. (MCRP 5-12C)

Marine expeditionary force (Forward)—A designated lead echelon of a Marine expeditionary force (MEF), task-organized to meet the requirements of a specific situation. A Marine expeditionary force (Forward) varies in size and composition, and it may be commanded by the MEF commander personally or by another designated commander. It may be tasked with preparing for the subsequent arrival of the rest of the MEF/joint/multinational forces, and/or the conduct of other specified tasks, at the discretion of the MEF commander. A Marine expeditionary force (Forward) may also be a stand-alone Marine air-ground task force (MAGTF), task-organized for a mission in which a MEF is not required. In a joint or multinational environment, it may also contain other Service or multinational forces assigned or attached to the MAGTF. Also called **MEF (Fwd)**. (MCRP 5-12C)

Marine expeditionary unit—A Marine air-ground task force (MAGTF) that is constructed around an infantry battalion reinforced, a composite squadron reinforced, and a task-organized logistics combat element. It normally fulfills Marine Corps' forward sea-based deployment requirements. The Marine expeditionary unit provides an immediate reaction capability for crisis response and is capable of limited combat operations. In a joint or multinational environment, it may contain other Service or multinational forces assigned or attached to the MAGTF. Also called **MEU**. (MCRP 5-12C)

Marine logistics group—The logistics combat element (LCE) of the Marine expeditionary force (MEF). It is a permanently organized command tasked with providing combat service support beyond the organic capabilities of supported units of the MEF. The Marine logistics group (MLG) is normally structured with direct and general support units, which are organized to support a MEF possessing one Marine division and one Marine aircraft wing. The MLG may also provide smaller task organized LCEs to support Marine air-ground task forces smaller than a MEF. Also called **MLG**. (MCRP 5-12C)

maritime prepositioning ship—(See JP 1-02, maritime pre-positioning ships, for core definition. Marine Corps amplification follows.) A maritime prepositioning ship is normally designated as a T-AKR. (MCRP 5-12C)

materials handling equipment—Equipment used at air, ground, and sea ports to handle large cargo. Also called **MHE**. (JP 1-02)

medical treatment facility—A facility established for the purpose of furnishing medical and/or dental care to eligible individuals. (JP 1-02)

mobility—A quality or capability of military forces which permits them to move from place to place while retaining the ability to fulfill their primary mission. (JP 1-02)

mortuary affairs—Provides for the search for, recovery, identification, preparation, and disposition of human remains for whom the Services are responsible by status and executive order. (JP 1-02)

movement control—The planning, routing, scheduling, and control of personnel and cargo movements over lines of communications, includes maintaining in-transit visibility of forces and material through the deployment and/or redeployment process. (JP 1-02)

N

naval beach group—A permanently organized naval command within an amphibious force comprised of a commander and staff, a beachmaster unit, an amphibious construction battalion, and an assault craft units, designed to provide an administrative group from which required naval tactical components may be made available to the attach force commander and to the amphibious landing force commander.(JP 1-02)

O

objective—1. The clearly defined, decisive, and attainable goal toward which every operation is directed. 2. The specific target of the action taken which is essential to the commander's plan. (JP 1-02)

obstacle—Any natural or manmade obstruction designed or employed to disrupt, fix, turn, or block the movement of an opposing force, and to impose additional losses in personnel, time, and equipment on the opposing force. (JP 1-02)

operating forces—Those forces whose primary missions are to participate in combat and the integral supporting elements thereof. (Proposed for inclusion in MCRP 5-12C)

operation—1. A sequence of tactical actions with a common purpose or unifying team. 2. A military action or the carrying out of a strategic, operational, tactical, service, training, or administrative military mission. (JP 1-02)

operational level of war—The level of war at which campaigns and major operations are planned, conducted, and sustained to achieve strategic objectives within theaters or other operational areas. See also **strategic level of war; tactical level of war**. (JP 1-02)

operation order—A directive issued by a commander to subordinate commanders for the purpose of effecting the coordinated execution of an operation. Also called **OPORD.** (JP 1-02)

operation plan—1. Any plan for the conduct of military operations prepared in response to actual and potential contingencies. 2. A complete and detailed joint plan containing a full description of the concept of operations, all annexes applicable to the plan, and a time-phased force and deployment data. Also called **OPLAN**. (JP 1-02)

operations center—The facility or location on an installation, base, or facility used by the commander to command, control, and coordinate all operational activities. (JP 1-02)

organizational maintenance—That maintenance that is the responsibility of and performed by a using organization on its assigned equipment. (JP 1-02)

P

port—A place at which ships may discharge or receive their cargoes. It includes any port accessible to ships on the seacoast, navigable rivers or inland waterways. The term "ports" should not be used in conjunction with air facilities which are designated as aerial ports, airports, etc. (NTRP 1-02)

procurement—The process of obtaining personnel, services, supplies, and equipment. (NTRP 1-02)

R

rear area—That area extending forward from a command's rear boundary to the rear of the area assigned to the command's subordinate units. This area is to provide primarily for the performance of combat service support functions. (MCRP 5-12C)

rear area security—The measures taken before, during, and/or after an enemy airborne attack, sabotage action, infiltration, guerrilla action, and/or initiation of psychological or propaganda warfare to minimize the effects thereof. (MCRP 5-12C)

rebuild—The restoration of an item to a standard as nearly as possible to its original condition in appearance, performance, and life expectancy. (NTRP 1-02)

recovery—Actions taken to extricate damages or disabled equipment for return to friendly control or repair at another location. (Part four of a four-part definition.) (JP 1-02)

repair—The restoration of an item to serviceable condition through correction of a specific failure or unserviceable condition. (NTRP 1-02)

repair and replenishment point—A combat service support installation, normally in forward areas near the supported unit, established to support a mechanized or other rapidly moving force. It may be either a prearranged point or a hastily selected point to rearm, refuel, or provide repair services to the supported force. (MCRP 5-12C)

repair cycle—The stages through which a repairable item passes from the time of its removal or replacement until it is reinstalled or placed in stock in a serviceable condition. (JP 1-02)

requisition—An authoritative demand or request especially for personnel, supplies, or services authorized but not made available without specific request. (Proposed for inclusion in MCRP 5-12C)

resupply—The act of replenishing stocks in order to maintain required levels of supply. (JP 1-02)

S

salvage—1. Property that has some value in excess of its basic material content but is in such condition that it has no reasonable prospect of use for any purpose as a unit and its repair or rehabilitation for use as a unit is clearly impractical. 2. The saving or rescuing of condemned, discarded, or abandoned property, and of materials contained therein for reuse, refabrication, or scrapping. (JP 1-02)

security—Measures taken by a military unit, activity or installation to protect itself against all acts designed to, or which may, impair its effectiveness. (Part one of a three-part definition.) (JP 1-02)

selective interchange—The controlled removal and replacement of a serviceable repair part or component from one item to satisfy a deficiency in another item. (Proposed for inclusion in MCRP 5-12C)

serial—An element or a group of elements within a series that is given a numerical or alphabetical designation for convenience in planning, scheduling, and control. (Part one of a two-part definition.) (JP 1-02)

ship-to-shore movement—That portion of the action phase of an amphibious operation that includes the deployment of the landing force from ships to designated landing areas. (JP 1-02)

stockage objective—The maximum quantities of materiel to be maintained on hand to sustain current operations, which will consist of the sum of stocks represented by the operating level and the safety level. (JP 1-02)

strategic level of war—The level of war at which a nation, often as a member of a group of nations, determines national or multinational (alliance or coalition) strategic security objectives and guidance, then develops and uses national resources to achieve those objectives. See also **operational level of war; tactical level of war**. (JP 1-02)

subordinate command—A command consisting of the commander and all those individuals, units, detachments, organizations, or installations that have been placed under the command by the authority establishing the subordinate command. (JP 1-02)

supply—The procurement, distribution, maintenance while in storage, and salvage of supplies, including the determination of kind and quantity of supplies. a. **producer phase**—That phase of military supply that extends from determination of procurement schedules to acceptance of finished supplies by the Services. b. **consumer phase**—That phase of military supply that extends from receipt of finished supplies by the Services through issue for use or consumption. (JP 1-02)

supply requirements—In logistic and combat service support those commodities that are essential to begin and sustain combat operations

support—1. The action of a force that aids, protects, complements, or sustains another force in accordance with a directive requiring such action. 2. A unit that helps another unit in battle. 3. An element of a command that assists, protects, or supplies other forces in combat. (JP 1-02)

supporting establishment—Those personnel, bases, and activities that support the Marine Corps operating forces. (MCRP 5-12C)

survivability—(See JP 1-02 for core definition. Marine Corps amplification follows.) The degree to which a system is able to avoid or withstand a manmade hostile environment without suffering an abortive impairment of its ability to accomplish its designated mission.

sustainment—The provision of logistics and personnel services required to maintain and prolong operations until successful mission accomplishment. (JP 1-02)

T

tactical level of war—The level of war at which battles and engagements are planned and executed to achieve military objectives assigned to tactical units or task forces. See also **operational level of war; strategic level of war**. (JP 1-02)

tactical-logistical group—Representatives designated by troop commanders to assist Navy control officers aboard control ships in the ship-to-shore movement of troops, equipment, and supplies. Also called **TACLOG group.** (JP 1-02)

task force—A component of a fleet organized by the commander of a task fleet or higher authority for the accomplishment of a specific task or tasks. (JP 1-02)

task organization— (See JP 1-02 for core definition. Marine Corps amplification follows.) A temporary grouping of forces designed to accomplish a particular mission. Task organization involves the distribution of available assets to subordinate control headquarters by attachment or by placing assets in direct support or under operational control of the subordinate. (MCRP 5-12C)

terminal operations—The reception, processing, and staging of passengers; the receipt, transit, storage, and marshalling of cargo; the loading and unloading of modes of transport conveyances; and the manifesting and forwarding of cargo and passengers to destination. (MCRP 5-12C)

throughput—(See JP 1-02 for core definition. Marine Corps amplification follows.) In logistics, the flow of sustainability assets in support of military operations, at all levels of war, from point of origin to point of use. It involves the movement of personnel and materiel over lines of communications using established pipelines and distribution systems. (MCRP 5-12C)

traffic management—The direction, control, and supervision of all functions incident to the procurement and use of freight and passenger transportation services. (JP 1-02)

U

unified command—A command with a broad continuing mission under a single commander and composed of significant assigned components of two or more Military Departments that is established and so designated by the President, through the Secretary of Defense with the advice and assistance of the Chairman of the Joint Chiefs of Staff. (JP 1-02)

Section III. Nomenclature

AH-1W	attack helicopter (Super Cobra)
AV-8B	attack aircraft (Harrier)
CH-53D/E	medium/heavy assault support helicopter (Sea/Super Stallion)
F/A-18	fighter/attack aircraft (Hornet)
MV-22	medium lift, vertical takeoff and tiltrotor aircraft (Osprey)
T-AH	hospital ship (MSC)
T-AVB	aviation logistics support ship (MSC)
V/STOL	vertical and/or short takeoff and landing aircraft
VTOL	vertical takeoff and landing

REFERENCES AND RELATED PUBLICATIONS

Department of Defense (DOD) Publication

4500.9-R   Defense Transportation Regulation, Parts I, II, and III

Department of Defense Instruction (DODI)
3020.41   Operational Contract Support (OCS)

Department of Defense Directive (DODD)
1300.22   Mortuary Affairs Policy
2010.9   Acquisition and Cross-Servicing Agreements

Allied Joint Doctrine Publication (AJP)

-4 (B) Allied Joint Doctrine for Logistics

Joint Publications (JPs)

1-02	Department of Defense Dictionary of Military and Associated Terms
2-0	Joint Intelligence
3-0	Joint Operations
3-02	Amphibious Operations
3-02.1	Amphibious Embarkation and Debarkation
3-05	Special Operations
3-07	Stability Operations
3-07.2	Antiterrorism
3-07.3	Peace Operations
3-07.4	Joint Counterdrug Operations
3-10	Joint Security Operations in Theater
3-34	Joint Engineering Operations
3-57	Civil-Military Operations
4-0	Joint Logistic
4-01	Joint Doctrine for the Defense Transportation System
4-01.2	Sealift Support to Joint Operations
4-01.6	Joint Logistics Over the Shore (JLOTS)
4-02	Health Service Support
4-06	Mortuary Affairs

Chairman of the Joint Chiefs of Staff Manual (CJCSMs)

3122.05	Operating Procedures for Joint Operation Planning and Execution System (JOPES)
3150.05D	Joint Reporting System Situation Monitoring Manual

MCTP 3-40B. Tactical-Level Logistics

Chairman of the Joint Chiefs of Staff Instruction (CJCSI)

3401.02B   Force Readiness Reporting

Army Techniques and Publications (ATP)
4-16   Movement Control

Marine Corps Publications

Marine Corps Doctrinal Publication (MCDPs)

1	Warfighting
1-0	Marine Corps Operations
1-1	Strategy
1-2	Campaigning
1-3	Tactics
2	Intelligence
3	Expeditionary Operations
4	Logistics
5	Planning
6	Command and Control

Marine Corps Warfighting Publications (MCWPs)

3-40.3	MAGTF Communication Systems
3-40.8	Componency
2-1	Intelligence Operations
3-17.7	General Engineering
3-21.1	Aviation Ground Support
3-21.2	Aviation Logistics
3-24	Assault Support
3-31.5	Ship-To-Shore Movement
4-1	Logistics Operations
4-2	Naval Logistics
4-11.1	Health Service Support Operations
4-11.3	Transportation Operations
4-11.4	Maintenance Operations
4-11.5	Seabee Operations in the MAGTF
4-11.7	MAGTF Supply Operations
4-11.8	Services in an Expeditionary Environment
5-1	Marine Corps Planning Process
6-22	Communications and Information Systems

Marine Corps Reference Publications (MCRPs)

3-33.1A	Civil Affairs Tactics, Techniques, and Procedures
4-11B	Environmental Considerations
4-11E	Contingency Contracting (12 Feb 09)
4-11H	*Multi-Service Tactics, Techniques, and Procedures for Operational Contract Support*
4-11.1G	Patient Movement
4-11.3D	The Naval Beach Group
4-11.3F	Convoy Operations
4-11.3H	MTTP for Tactical Convoy Operations
4-11.8A	Marine Corps Field Feeding Program
5-11.1A	MAGTF Aviation Planning Documents
5-12C	Marine Corps Supplement to the Department of Defense Dictionary of Military and Associated Terms
5-12D	Organization of the United States Marine Corps
6-12C	Commander's Handbook for Religious Ministry Support

MAGTF Staff Training Pamphlet (MSTP)

4-0.2	Logistics Planner's Guide

Naval Doctrine Publications (NDPs)

4	Naval Logistics
5	Naval Planning
6	Naval Command and Control

Naval Tactics, Techniques and Procedures (NTTPs)

3-02.1M	Ship-to-Shore Movement
3-02.1.3	Amphibious/Expeditionary Operations Air Control
3-02.3	Maritime Prepositioning Force (MPF) Operations
4-02	Medical Logistics
4-02.2M	Patient Movement
4-02.4	Expeditionary Medical Facilities

Secretary of the Navy Instruction (SECNAVINSTs)

4000.37A	Naval Logistics Integration
1-1	Strategy
1-2	Campaigning

MCTP 3-40B. Tactical-Level Logistics

To Our Readers

Changes: Readers of this publication are encouraged to submit suggestions and changes that will improve it. Recommendations may be sent directly to Commanding Officer, Marine Corps Logistics Operations Group. Marine Air ground task force Training Command, Marine Corps Air Ground Combat Center, Box 788400, Twenty-nine Palms California 92278-8400, fax to 760-830-3767 (DSN 830-3767)

Recommendations should include the following information:

- Location of change
    - Publication number and title
    - Current page number
    - Paragraph number (if applicable)
    - Figure or table number (if applicable)
- Nature of change
    - Add, delete
    - Proposed new text, preferably double-spaced and typewritten
- Justification and/or source of change

Additional copies: An electronic copy may be obtained from the Doctrine Division, MCCDC, World Wide Web home page, found at: **http://www.doctrine.usmc.mil**.

Unless otherwise stated, whenever the masculine pronouns are used, both men and women are included.

Made in the USA
San Bernardino, CA
08 May 2017